COMMON HOUSEHOLD PESTS

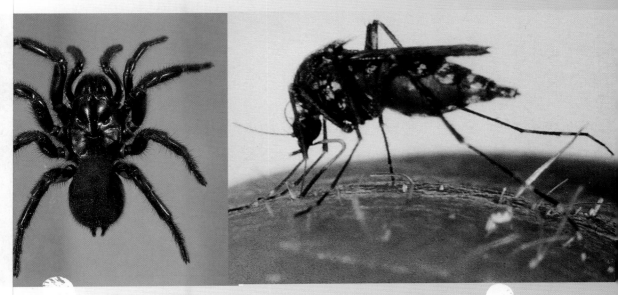

A HOMEOWNER'S GUIDE TO DETECTION AND CONTROL

Phillip HADLINGTON & Christine MARSDEN

UNSW PRESS

CONTENTS

A UNSW PRESS BOOK

Published by
University of New South Wales Press Ltd
University of New South Wales
Sydney 2052 Australia
www.unswpress.com.au

© P. Hadlington and C. Marsden 1999

First published 1999

National Library of Australia
Cataloguing-in-Publication entry:

Hadlington, Phillip W., 1923– .
Common household pests: a homeowner's guide to detection and
control.

Bibliography.
Includes index.
ISBN 0 86840 625 2.

1. Household pests — Control — Australia.
I. Marsden, Christine.
II. Title.

648.70994

Designer Di Quick + Dana Lundmark
Printer Everbest Printing, Hong Kong

FOREWORD

A year or so ago, when writing the foreword to the companion book *Termites and Borers: A Homeowner's Guide to Detection and Control*, I observed that the book would help to avoid or reduce misunderstandings brought about by lack of knowledge. It did and continues to do so!

That success has led to the production of this book by the same authors. Like its predecessor, rather than being a do-it-yourself pest control instruction manual, it fills an important gap by providing sufficient information, in simple terms, for home-owners to understand the potential hazards of household pests.

I recall an occasion some 25 years ago when Phil Hadlington considered the factors which affected the reproductive potential of organisms. These factors included climatic conditions, the parasites, the predators and the food supply. He maintained that the 'backyard ecology' in which the organisms lived was important to the understanding of the behaviour of household pests.

That same philosophy permeates this very useful book. Together with Christine Marsden, Phil Hadlington has, once again, made a significant contribution to 'de-mystifying' those organisms with which householders often share their environment.

Doug Howick
National Executive Director
Australian Environmental Pest Managers Association

INTRODUCTION

Most people in Australia have experienced problems which are due to pests in their homes. In many cases these pests are tolerated because homeowners are unaware that there are both environmental and chemical methods for the control of these unwanted creatures. When John Elliot of University of New South Wales Press asked us to prepare a book on the subject to assist the householder we realised there was a need for such a publication, and were pleased to produce this book.

Household Pests: A Homeowner's Guide to Detection and Control has been prepared to assist the householder to detect, control and prevent infestations of pests in the house. Many pests such as house flies and mosquitoes can be controlled within certain limits by the householder, but other pests may require regular attention by the pest control technician.

Termites, borers and agents of decay are the most destructive household pests, but these are not covered in this book. They receive special attention in our companion book *Termites and Borers: A Homeowner's Guide to Detection and Control*.

The initial step in pest control is an accurate identification of the cause, and there are many photographs and illustrations to assist the householder in this identification. Once the pest is known the householder will then be able to determine whether the service of a pest control technician is required.

For those householders concerned about the application of pesticides, there is a chapter on environmental control. For many pests, such as food and fabric pests, certain environmental procedures can maintain a clear house, if carried out regularly. Environmental measures also complement pesticide controls, so that the number of insects such as cockroaches, fleas and ants may be kept at tolerable levels for longer periods.

The tables in Chapter 15 provide details on the pesticides and the specific material and formulations needed to control the pests. They also assist the householder in discussion with the pest control technician in selecting a pesticide for a particular treatment.

The importance of the details on the pesticide label is stressed in the chapters on pesticides and chemical control. If the label data for particular formulations of the same pesticide do not state the pest or group of pests the formulation is not suitable for that pest.

Acknowledgments

The illustrations of several insects and their life cycles were prepared by Noel Cooney. These illustrations complement the photographs which appear throughout this book. Some of the insects are very small and Noel's illustrations show features which are visible only with a microscope.

We are also grateful to Gary Beehag of Globe Chemicals Pty Ltd who provided us with the trade names and active constituents of many products available to the pest-control industry and the householder.

COCKROACHES

One of the most commonly encountered cockroaches in houses is the German cockroach (from left: nymph, adult male and female, and egg case). The adult is 12–15 mm long and winged, but it does not fly. Note the two dark, longitudinal bands on the thorax.

Cockroaches have been present on earth for over 300 million years. They are the most common pests of houses and restaurants throughout Australia. Over the years certain species of cockroaches have co-habited with humans in many areas such as homes, ships, food production and storage facilities. Some species of cockroaches inhabit sewer systems, drains and composts used for the garden. Contamination of food and equipment has occurred. Some pest species of cockroaches have contact with animal and human faeces as well as contaminated food, and they transfer these organisms on their bodies and in their faeces to food which is being prepared for human consumption. Many human diseases have been found in and on the bodies of cockroaches. These include organisms which cause food poisoning, gastroenteritis, dysentery, hepatitis and tuberculosis.

The pest species of cockroaches eat a wide range of foods so that most domestic situations are attractive to them. Crumbs, dried liquids on carpets, hair and even starchy bindings of books ensure their survival. While some species of cockroaches can fly, most dispersal occurs through the transport of food, food cartons and other goods.

LIFE CYCLE & HABITS

Cockroach eggs are encapsulated in a purse-shaped egg case which, depending on the species, contains between 12 and 40 eggs. The egg cases are either dropped in areas where the cockroach is active, or in some species are glued to surfaces. Cockroaches hatch as nymphs, which have a close association with others of the same species, for they have a natural gregarious behaviour. The young grow through several 'moults', in which they shed their cuticles, until they are adults. Cockroaches have no pupal or resting stage and the nymph's wings also gradually develop at each moult until they are, in most pest species, fully winged adults.

Some species reach maturity within three to four months, while other species require about 12 months. Depending on the species their life span varies from a few months to over one year. During that time from 5–30 egg cases could have been produced.

The pest species consume a wide range of foods returning to their hiding places during the day; usually in cracks and crevices in walls and even in furniture. It is important that computers and microwave ovens be protected for some cockroaches nest in these warm pieces of equipment, often resulting in expensive repairs to the electronic components.

Cockroaches are nocturnal, hiding during the day. Homeowners usually do not realise the magnitude of a problem unless the cockroach activity is viewed during the night. Because of the cockroaches' gregarious behaviour, sites of infestation can contain hundreds of both developing and mature cockroaches.

One feature of cockroach behaviour is that they groom themselves by passing their antennae and legs through their mouthparts. This practice has been exploited in the application of some chemicals used in their control.

7

COCKROACH SPECIES

Australia has over 400 native species of cockroaches which are of no concern to the homeowner. There are six pest species, which apparently had their origin in tropical and subtropical Africa and now occur over most of the world. These pest species were introduced into Australia over the last 200 years and are the most often encountered household pests in Australia.

German cockroach
Blattella germanica

This is one of the smaller cockroaches, amber-brown in colour and with two longitudinal dark stripes on the thorax. It is the most prolific breeder of the pest cockroaches, having three to four generations a year and taking only 40 days to mature from egg to adult during the summer.

German cockroaches are usually found in kitchens behind and under stoves, dishwashers and sinks. During recent years microwave ovens and computers have had their electronic controls damaged by cockroaches that have been attracted to the warmth. When German cockroaches are seen in rooms other than the kitchen, it is likely that the population of the insects is very dense throughout a building.

American cockroach
Periplaneta americana

The adult American cockroach is 35–40 mm long with a pale yellow border around the upper area of the thorax. It is capable of flight. The female may lay up to 50 egg cases, each containing 10–25 eggs. The egg case is often glued to surfaces usually near food. From egg to adult takes 6–12 months, and the adult can live for at least six months.

American cockroaches live both indoors and outside depending on conditions. They inhabit wall

Adult American cockroaches are 35–40 mm long with a pale yellow border around the dorsal area of the thorax. They are common in homes especially where food is present.

cavities, roof areas and subfloors and are often encountered around drains and sewers. They are therefore potential spreaders of disease organisms.

Australian cockroach
Periplaneta australasiae

The adult Australian cockroach is 30–35 mm long. It has a dark body with pale yellow areas on the upper area of the thorax and on the fore margins of the front wings. The female may produce 20 egg cases each containing 20–24 eggs during her lifetime.

This species favours warmer climates and occurs mostly outside, often where there are plants, wood piles and under garden litter. They also occur in roof voids and walls.

Brownbanded cockroach
Supella longipalpa

This is a smaller cockroach, only about 10–14 mm long. Males have full wings, but females only short wings. This species is identified from others by pale bands across both the thorax and abdomen. Their egg cases, which may contain up to 18 eggs, are usually glued to surfaces. They have a short development period of from two to four months. Adults usually live for about six months.

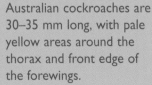

Australian cockroaches are 30–35 mm long, with pale yellow areas around the thorax and front edge of the forewings.

Brownbanded cockroaches are about 10–14 mm as adults, with pale bands across the body. Note the egg case at the end of the abdomen of this female.

The adult smokybrown cockroach (right, with nymph) is about 30–35 mm and dark brown with full wings. They are often attracted to lights.

Brownbanded cockroaches occur mainly inside buildings where they occupy wardrobes and office furniture. They are also often encountered in hospitals and restaurants.

Smokybrown cockroach
Periplaneta fuliginosa

This is a large cockroach of a uniform dark brown colour, measuring 30–35 mm long. Adults live for 6–12 months, and are fully winged and capable of flight. During its life the female may lay 20 egg cases, each containing about 26 eggs.

This species is found in roof voids of houses, walls and subfloor areas. They enter houses from garden areas and greenhouses where plant materials form much of their food. They are not often encountered in kitchens and other rooms of the house.

Oriental cockroach
Blatta orientalis

This is a medium-sized uniformly brown cockroach, about 20–25 mm long, with reduced wings in both the male and female. One female may produce about 8–14 egg cases during her life, each containing 12–18 eggs.

The oriental cockroach is found in cooler climates. It lives under litter and garden mulch where vegetation decay occurs. It also frequents garbage disposal areas. It can also be found in damp areas of cellars, subfloors and walls.

The reduction of populations of pest cockroaches is achieved by the implementation of methods of sanitation and hygiene as well as by the application of chemicals. These two methods of control should complement each other.

DETECTION & PREVENTION

Like most animals cockroaches need food, water and shelter. These three requirements are present in most houses and buildings where humans live and work. The following preventative measures are recommended:

1   Eliminate food and water sources. This is achieved by implementing clean-up procedures at the end of each day so that food is not available at night when cockroaches are active. This is particularly relevant in offices, factories and food shops as well as houses.

2   Store food in sealed containers and in refrigerators to reduce accessibility by cockroaches.

3   Keep garbage in sealed containers prior to collection.

4   Inspect in-coming containers from food stores prior to unwrapping within the house or business.

5   Heat treat appropriate food articles in a microwave to kill egg cases before storage in the kitchen.

6   Regularly inspect the motors of washing machines and refrigerators, and the areas under and behind stoves, as these provide cockroaches with shelter and warmth.

7   Monitor microwave ovens, fax machines and computers, as these also provide shelter and warmth particularly in winter. This may help avoid expensive repairs.

8   Use vacuum cleaners to remove food waste from floors and cupboards in kitchens. Wiping may leave residues in cracks.

9   Fill in cracks and crevices in walls and cupboards to prevent cockroach access.

10 Inspect roof cavities, as this can often locate populations of otherwise undetected smokybrown and American cockroaches.

Oriental cockroaches are about 20–25 mm long and a uniform dark brown. They mostly live near ground level but do enter houses and feed on starches in book bindings and food residue.

11

CONTROL

Chemical control

The choice and application of insecticides depends on many factors such as the species of cockroach, their numbers, location, and the sensitivity of human occupants to chemicals. An inspection of the entire property including the roof cavity, subfloor and the immediate garden area must precede any chemical application.

Several groups of insecticides are approved for the control of cockroaches. These include organophosphorus insecticides such as chlorpyrifos, azamethiphos, diazinon, fenthion and dichlorvos. Carbamate insecticides such as propoxur and bendiocarb are also effective. The synthetic pyrethroids such as permethrin and deltamethrin have also proved effective against most species of cockroach.

During recent years a group of insecticides known as 'insect growth regulators' (IGRs) have been developed and these are effective against most species of cockroach. These pesticides disrupt the normal function of the insects to cause death. There are two types of IGR: juvenile hormone analogues which disrupt the normal development of cockroaches; and chitin synthesis inhibitors which disrupt the normal formation of the cuticle (causing death during moulting as the cockroach cannot replace its shed cuticle. Control using these materials may occur from three to five months after application, so that it is often necessary to add a 'traditional' insecticide such as permethrin to kill the adults while the IGR is there to affect and finally kill off the developing nymphal stages.

IGRs have a very low toxicity to humans, but have a long-term residual effect on the cockroach populations. They include fenoxycarb, hydroprene, methoprene and triflumuron. These are available as aerosols and are applied to areas such as cracks and crevices where cockroaches hide during the daytime.

Dusts containing boric acid are also available and are applied directly to cracks and crevices. These dusts are oral poisons, and are ingested by the cockroach during grooming.

Chemical baits have also proved effective in controlling cockroaches, particularly the German cockroach. Baits are often formulated in gels and can be placed in strategic situations in and under kitchen cupboards and near computers. Some contain either hydramethylnon and chlorpyrifos, which have a low toxicity to mammals. Baits should be checked and changed regularly. Chlorpyrifos is also available in a lacquer formulation which can be applied to various surfaces.

While most insecticides such as chlorpyrifos and deltamethrin are effective for some months, the growth regulators and baits complement these sprays giving long-term control.

Fipronil, marketed as Goliath Cockroach Gel, is a recently developed material which is formulated in a gel and applied as drops to carefully selected sites. Cockroaches are attracted to the gel and feed upon it. It is also taken back to their hiding places where other cockroaches consume it and die soon afterwards. This bait can only be applied by a licensed pest technician.

Sticky traps may also be used to determine the presence of cockroaches and to give an indication of the numbers present. They do not usually achieve adequate control on their own.

The most favoured areas for cockroaches are kitchens bathrooms and for some species roof cavities. Special attention to areas surrounding microwave ovens, dish washers, stoves and refrigerators gives best results. While pest control technicians carry out their treatments during the day, night-time applications are best done after dinner, when cockroaches are most active.

Blocks of flats, residential units and semi-detached houses usually require treatment together for the best long-term results, which may require organisation through bodies corporate, estate agents or residents' associations.

COMMON QUESTIONS ABOUT
COCKROACHES

Q Why is it important to have the cockroach identified?

A Cockroach species differ in their habits and habitat. These may determine the type and location of treatment. The cockroach may be a non-pest species from the garden, requiring only minor treatment.

Q Who can I contact to identify the species?

A Pest control technicians will provide this service, as will some governmental scientists or consultants.

Q If I do not want insecticides such as organophosphorus, carbamate or pyrethroids used, are there other options?

A Yes, the insect growth regulators have been developed to assist in this problem. They have a very low toxicity, but do take some months to kill most of the cockroaches present.

Q Why is a vacuum cleaner recommended for kitchen use and for cupboards?

A Cockroaches survive on small amounts of food which may be left after sweeping with a broom or wiping cupboards with a cloth. A small vacuum cleaner leaves very little residue when used after the evening meal and before cockroaches become active at night.

Q When should a treatment by a pest control operator be done?

A The best time for an initial treatment is early to mid-spring, but this depends on the magnitude of the problem.

Q Have cockroaches any natural parasites and predators?

A Yes, certain wasps lay their eggs in cockroach bodies, and birds such as magpies and currawongs, lizards and spiders will locate and eat them. However, these are of no significance within a house or restaurant.

Q Are cockroaches carriers of disease organisms?

A Certain species of cockroach are physical carriers of disease organisms, for they are found in and around sewer systems, garbage areas, animal faeces on the ground and garden manures.

CHAPTER 2

FLEAS

Fleas can be pests in homes, hotels, motels and wherever human populations occur, often where these areas are shared with animals. There are many species of fleas, some having particular preferences for hosts. All have piercing and sucking mouthparts used to draw blood from their host. In so doing they can transmit diseases to humans and animals, and are therefore important pests both in the medical and veterinary areas.

An understanding of the habits and life cycles of fleas is important in their prevention and control.

APPEARANCE, LIFE CYCLE & HABITS

Adult fleas vary in size, but are usually about 1–2 mm long, brown in colour and compressed laterally (that is they have very thin bodies) so that they can move freely between the hairs of their animal hosts. Larvae are tiny and legless, with hairs on their body. They have chewing mouthparts and feed on food particles and blood faecal pieces from adult fleas.

The adult female may lay some hundreds of eggs in her lifetime, but usually about four to eight eggs after each blood meal. The eggs are laid on the animal and these fall off onto floors, the ground and often in its sleeping and resting area. The eggs hatch in from four to ten days. The areas surrounding a house can also be a significant source of fleas.

The larvae feed on many forms of organic matter located in carpets and other floor coverings as well as outside in lawns. When finished feeding and fully grown, usually after about 12–20 days, the larvae pupate in silken cocoons which are often covered with debris. After about seven to ten days the adults emerge and seek a blood meal from the host. The larval and pupal stages, as well as the adult, may extend for many months depending on the availability of food.

Warm and humid conditions such as those which occur in summer and autumn favour the development of flea larvae, pupae and adults. Flea pupae may remain in floor coverings and cracks in flooring for several months and often the adults emerge from their cases when vibration occurs. This happens when a house is vacated during holidays, mostly at Christmas, and when the family returns the vibration from walking fractures the pupal cases. The adult fleas then seek blood meals from the legs of those nearby!

FLEA LIFE CYCLE
THE HUMAN FLEA *PULEX IRRITANS*

adult 1.0–1.5 mm

body laterally thin

piercing sucking mouthparts

well developed coxa for jumping

variable periods a week to some months

Pupa

in small silken cocoon. vibration causes it to break

Eggs - 0.5 mm
laid in fur and hairs and clothing
falls from animals

4–8 days to hatch

small bristles pointing backwards

mandibulate mouthparts

larva 1.0–2 mm

legless and white
present in cracks and carpet. Eat particles
of food, faeces and organic debris

16

FLEA SPECIES

The main pest fleas in domestic situations are:

Dog flea
Ctenocephalides canis

This flea is similar to other fleas and has a wide range of hosts, but has been encountered much less during recent years.

Cat flea
Ctenocephalides felis

The cat flea also has a wide host range feeding on cats, dogs, humans and other animals. It is the most often encountered species in much of Australia at present.

Oriental rat flea
Xenopsylla cheopis

The roof rat is the preferred host of the rat flea.

Human flea
Pulex irritans

This flea also attacks animals such as dogs, cats, pigs as well as humans, but the incidence of attack is less now than in previous years.

FLEAS & DISEASE

Fleas are one of the most significant conveyers of diseases to humans and also some animals. Over the last one thousand years countless millions of people have died because of this transmission of diseases by fleas.

Bubonic plague
'The Black Death'

This flea-borne disease caused the deaths of millions during the Middle Ages. The disease organism is the bacterium *Yersinia pestis* which is usually carried by rats, mostly the roof rat. It is passed on from one rat to another by fleas and then to humans. The flea species which is the main vector of the plague is the oriental rat flea (*Xenopsylla cheopis*). Australia has not been troubled with this disease for at least fifty years.

Murine or endemic typhus

This disease is also transmitted mainly by the oriental rat flea. The micro-organism which causes this disease

17

condition is *Rickettsia typhi*. When the flea bites a human it leaves faeces behind, carrying the micro-organism. When the bite is scratched, the infected faeces of the flea contaminates the wound.

Intestinal parasites

The dog tapeworm *Dipylidium caninum* is conveyed between dogs by fleas. Rodent tapeworm is passed on to other rodents when the fleas are infected. Children have occasionally been infested by these tapeworms.

Flea allergy dermatitis

Fleas produce a severe form of allergic dermatitis in dogs and cats. Some animals are more susceptible to this form of dermatitis than others. Flea control is important in the prevention of this condition.

CONTROL

The householder may successfully control fleas even if animal pets share the living area. Control falls into two categories, one complementing the other: non-chemical preventive methods; and chemical methods.

Non-chemical preventive methods

The breeding areas for fleas from animal pets can be significantly reduced by washing floors regularly and vacuuming carpets, including under furniture and areas where animals may rest. Animal beddings should be removed daily and shaken well away from the home, or left in the sun to remove eggs.

The contents of the vacuum cleaner should be heat-treated in a black plastic bag — by placing it in the sun for a few hours. It may also be treated with an aerosol spray before being emptied. The vacuum cleaner and bag, must not be stored in a partially full or full condition as the environment created inside the bag is ideal for flea breeding.

A thorough vacuuming is important to reduce any flea population present before going on holidays.

Chemical controls

When there are pets in a house, the animal should be treated at the same time as the house. Treatment of animals is usually done with products specifically registered for that purpose. Generally these are either

powders (such as permethrin, carbaryl and synergised pyrethrins) or washes (of maldison, carbaryl, permethrin and synergised pyrethrins). Insecticidal collars on animals may be used to obtain long-term control. The chemicals used to impregnate the collars are dichlorvos, permethrin and chlorpyrifos.

Systemic insecticides, such as cythioate, may be administered to pets. Fleas die when consuming blood containing the insecticide. Products such as fipronil can be used, and 'Frontline' and 'Program' are available in tablet form to protect dogs from fleas. Such insecticides must be prescribed by a veterinary surgeon.

Surface sprays on floors, carpets and verandahs outside a house have a residual and effective life of two to three months depending on the situations. Surface sprays include organophosphorus insecticides (chlorpyrifos), carbamates (bendiocarb) and pyrethroids (deltamethrin and permethrin). Special attention to cracks, crevices and wall joints is usually recommended. Subfloor and some outside areas may also need attention.

An adult flea with well developed rear legs for jumping.
It has piercing and sucking mouthparts, and has a flattened body to pass between hairs.

The use of IGR sprays can disrupt the development of fleas and their reproduction for some months. These have a very low toxicity to animals and include mat-erials such as triflumuron, pyriproxifen, fenoxy-carb and methoprene. IGRs used on their own achieve control in the long term, but to achieve an immediate and quick reduction permethrin in added to several formulations.

Space spraying using either pyrethrins or dichlor-vos can be used only when the premises are vacated for at least a few hours and achieve a rapid kill of adult fleas. IGRs such as fenoxycarb and methoprene are applied as aerosols or 'aerosol bombs'. Dusting using permethrin or bendiocarb can be done where sprays cannot be applied.

FLEAS

WHAT TO DO

Sites where fleas occur are different, and the treatments may vary. However, following the general procedure in the sequence below is suggested.

1   Carefully vacuum the entire house.

2   Heat-treat the vacuum bag in the sun. Never put the cleaner away with anything remaining in the bag, unless it has been treated.

3   Check any animals (especially cats and dogs) and either take them to a veterinary clinic, or wash and treat them with a pesticide registered for the purpose of flea removal.

4   Use a surface spray of a registered insecticide, preferably without odour and having a low toxicity to humans, to treat floor surfaces particularly where animals rest or sleep. (The pyrethroids permethrin and deltamethrin are suitable for surface treatments, as are the IGRs triflumuron and pyriproxifen, although these take longer.)

5   Vacuum the affected areas daily (or every second day) until the fleas are controlled. Remember to treat the contents of the bag after each use.

CHAPTER 3

ANTS

Ants are social insects which live in nests. They have several 'castes' with specific duties for the survival of the colony. For this reason they are often confused with termites ('white ants'), but the two groups have very little in common apart from their size and social behaviour.

Ants feed on a wide range of foods, from those found in homes to the sugary excretions from plant bugs. They are the most frequently encountered insects in and around the average home.

Carpenter ants often nest in decayed wood

BROWN ANTS

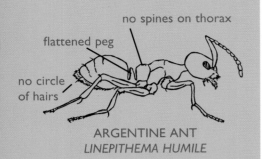

no spines on thorax
flattened peg
no circle of hairs

ARGENTINE ANT
LINEPITHEMA HUMILE

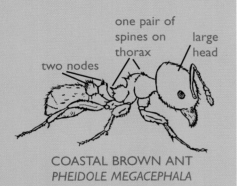

one pair of spines on thorax
large head
two nodes

COASTAL BROWN ANT
PHEIDOLE MEGACEPHALA

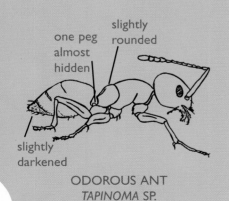

slightly rounded
one peg almost hidden
slightly darkened

ODOROUS ANT
TAPINOMA SP.

Ants are considered nuisance pests in the home as they enter from outside in the garden and make their way to the food handling facilities. Depending on the species of ant their nests are made under paths, in plant pots, wall cavities, subfloor areas and in roof cavities. Ants can also carry disease organisms. Certain species are attracted to dog faecal droppings, and to other waste products in garbage containing organisms which cause dysentery or pathogenic bacteria such as *Salmonella*.

FEATURES & LIFE CYCLE

Most species of ants have three distinct body segments (the head, thorax and abdomen), 'elbowed' antennae, and a constriction of body segments between the thorax and abdomen. If wings are present the forewings are larger than the hindwings.

There are four stages in the life cycle of ants: egg, larva, pupa and adult. The larvae which hatch from the eggs are fed by the adults, and the pupa is a resting stage before transforming into an adult. A colony consists of a queen, a king which dies soon after fertilising the queen, and workers. The workers which are sterile females, are the most abundant caste and their duties include defence, food collection, feeding the other castes and nest construction. Some workers with large heads are known as 'soldiers'.

The queen produces eggs which develop into the various castes. Swarms of winged mature male and female ants annually leave to establish new colonies, and also some females and workers may leave a parent colony from time to time to set up new colonies, referred to as 'budding off'.

The worker ants seek food and when located, the information is communicated to the rest of the colony. This is done in several ways, including trail-marking by pheromone secretions, and taste. Ants are also predatory, preying on other insects, particularly termites in the case of some ant species.

Many people say 'an ant has bitten me', but they are not technically correct. A bite is done with the jaws, but ants which cause pain usually do so with a sting.

ANT SPECIES

There are many species of ants which are pests of homes. Because they have differing food preferences, their identification is important to determine the control procedure, particularly when baiting is used and an attractive food is needed with the poison.

Argentine ant
Linepithema humile

These are brown ants about 1.5–3.0 mm long. They have no odour when crushed. Nests are mostly found outside, although they move indoors during wet weather. They prefer sweet food.

Coastal brown ant
Pheidole megacephala

These are light-brown ants about 1.5–2.5 mm long. They are often located in walls of houses and behind skirting. They prefer meat and fats rather than sweet foods.

Pharaoh's ant
Monomorium pharaonis

Pharaoh's ants are light brown, about 2–3 mm long, and have no odour when crushed. They nest in walls, ceilings and subfloor areas. They eat a wide range of foods, including meat, vegetables and sweets.

Singapore ant
Monomorium destructor

This is another light-brown ant, about 2–3 mm long, which nests in walls, under cupboards and in subfloor areas. It prefers foods of animal origin, but will feed on sweet materials.

Whitefooted house ant
Technomyrmex albipes

These are black ants about 2.5–3.0 mm long with pale feet. They nest in cavity walls, rockeries and behind kitchen cupboards.

Carpenter ants
Camponotus spp.

This group contains several species which range in colours from brown to pale brown and range in size

BROWN ANTS

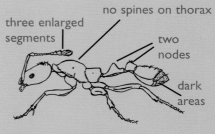

three enlarged segments — no spines on thorax — two nodes — dark areas

PHARAOH'S ANT
MONOMORIUM PHARAONIS

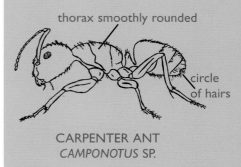

thorax smoothly rounded — circle of hairs

CARPENTER ANT
CAMPONOTUS SP.

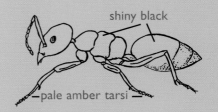

shiny black — pale amber tarsi

WHITEFOOTED HOUSE ANT
(BLACKANT)
TECHNOMYRMEX ALBIPES

23

from 7.0–12.0 mm. They mostly nest in decayed wood and eat live and dead insects as well as sweet foods.

Greenhead ants
Rhytidoponera spp.

These are black ants with metallic green heads. They are about 5.0–6.0 mm long. They have a characteristic odour and painful sting. They nest under paths and feed mainly on vegetable materials.

Meat ants
Iridomyrmex spp.

Meat ants are red and black, and about 13–14 mm long. Their nests are low mounds of soil with gravel on the surface. They eat mainly animal products and some sweet materials. These ants do not sting.

Bulldog ants
Myrmecia spp.

These species are either black or red, and range in length from 12-15 mm. Their nests occur mainly in bushland areas, where these are low mounds. All bull-dog ants can sting.

CONTROL

Non-chemical measures

Infestations can be reduced significantly by limiting food particles in rooms and cupboards of the house. Cleaning up food residues can also help eliminate ant problems. Food left outside for animals can attract ants which then go in search of more food, gaining entry into the house.

Chemical measures

Insecticides may be used as sprays applied to the nest or when this cannot be found, to the areas where the ants gain access to various parts of the house. When the nest is located the eradication of the colony is usually achieved. Some of the insecticides approved for this use are the organophosphates chlorpyrifos and diazinon, carbamates bendiocarb and propoxur, and the synthetic pyrethroids deltamethrin and cypermethrin.

Space sprays of dichlorvos or pyrethrins can be applied to the sites of activity, but better results are obtained if the nest is located. Chlorpyrifos when formulated as a lacquer has a long-term residual effect when applied to particular surfaces.

Temporary control is usually achieved by the use of insecticidal dusts such as permethrin or bendiocarb. Dusts may also be used as a special application in enclosed spaces such as electric power boxes and roof cavities.

Ant baits can be used successfully where insecticides can not be used or when the nest can not be located. The bait is collected at feeding sites, taken back to the colony by the workers, and fed to others including the queen and developing larvae. This usually results in the demise of the whole colony. The bait must be of a formulation which is attractive to the particular species. Hydramethylnon has proved successful in the eradication of the colony of certain species. Boron is also used in the preparation of baits for particular species. Baiting procedures are usually on the label. Granular formulations of ant baits are being used more in homes because of their ability to eliminate entire ant colonies.

ANTS
WHAT TO DO

When ants occur inside a house the following procedure is suggested:

1   Inspect the house and the area outside to locate the nest or the area from which most of the infestation originates.

2   Identify the species of ant, and determine the appropriate control measures. (Identification also often assists in the location of the nest.)

3   Treat the nest, or surfaces where ants are active, using either sprays or dusts.

4   Baits may be used where appropriate or where insecticides are not to be applied.

5   Eliminate food particles by vacuuming and other methods of hygiene both inside and outside the house.

25

CHAPTER 4

FLIES

Flies are important spreaders of disease organisms such as Salmonella food poisoning, dysentery, hepatitis, typhoid fever and various internal parasites. Some species, such as biting midges and March flies, have piercing and sucking mouthparts and inject a saliva which also causes skin irritation.

FEATURES & LIFE CYCLE

All flies belong to the insect order Diptera, which also includes mosquitoes. Adult flies have two wings, sucking or piercing and sucking mouthparts depending on the species, and a pair of compound eyes. The larvae are legless and often referred to as 'maggots'.

The adult fly lays its eggs, or in some cases produce live larvae, in various food materials which are often decomposing. The maggots, because of their respiration, create a slightly higher temperature which reduces their feeding time to several days or a week or so. When finished feeding they often leave the substance on which they have been feeding and pupate nearby. The pupal stage may last for days or weeks depending on the temperature. The adult flies can spread several kilometres from their larval breeding sites.

FLY SPECIES

House fly
Musca domestica

These are widely distributed and pests in homes, often because they also feed from waste in the garden. The adult fly has a 'sponging' type of sucking mouthpart to feed on liquid foods, and spreads disease organisms from one area to another. They are active during the day, inactive at night, and favour areas of higher temperatures.

The adult house fly is about 4–6 mm and lives for two to three weeks. The maggot' or larva is only about a week old before it pupates.

Bush fly
Musca vetustissima

The adult bush fly is about 5 mm long and is distinguished from the house fly by having two longitudinal bands on the thorax. They have both sponging and sucking mouthparts. They are outdoor pests, being attracted to humans as well as animals. They often cause eye conditions such as conjunctivitis.

Lesser house fly
Fannia canicularis

These have the habit of making short darts in flight, and also circling around animals and humans.

Stable fly
Stomoxys calcitrans

These resemble the house fly but have piercing and sucking mouthparts, and suck blood from humans and animals. They prefer outdoors and are pests in dairies and stables where they obtain the blood meals which are necessary prior to producing young.

Other commonly encountered flies are blowflies, flesh flies, vinegar flies, moth flies and biting midges.

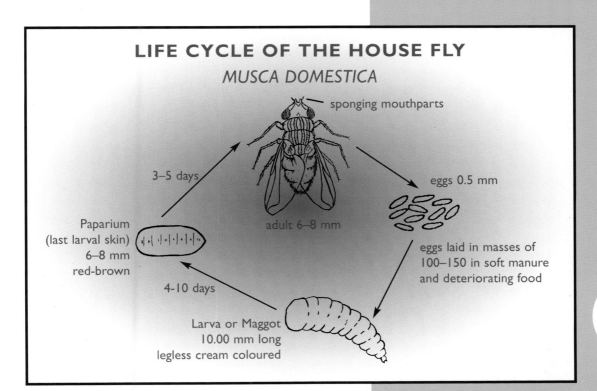

LIFE CYCLE OF THE HOUSE FLY
MUSCA DOMESTICA

sponging mouthparts

3–5 days

eggs 0.5 mm

Paparium
(last larval skin)
6–8 mm
red-brown

adult 6–8 mm

eggs laid in masses of
100–150 in soft manure
and deteriorating food

4-10 days

Larva or Maggot
10.00 mm long
legless cream coloured

Blowflies are commonly encountered pests of homes. They carry disease organisms.

The March fly is a blood-sucking fly, mostly encountered in bushland areas.

CONTROL

The most effective control involves the proper disposal of garbage. This will slowly reduce the fly populations by depriving them of breeding sites.

Trapping, adhesive flypaper, and electric fly traps which electrocute flies are all effective in reducing fly populations. Wire mesh screening of doors and windows is effective, and external ultraviolet electric fly 'zappers' are used particularly in areas where people gather for parties and entertainment, and commercial premises.

Flesh flies are attracted to decomposing foods.

There are several types of chemical control. The application instructions will appear on the label of the containers.

Fly repellents for personal use have particular application during the warmer months. These are used mainly for the bush fly and biting flies such as sand flies (midges), stable flies and March flies. Repellents are available as roll-ons, lotions and aerosols and these include chemicals such as diethyl toluamide and N-octyl bicycloheptene dicarboximide. The natural pyrethrins are also used. Repellents usually have an effective repellent life of from one to two hours.

Surface sprays are usually organophosphorus, carbamate and synthetic pyrethroid insecticides and when applied give an effective contact life of two to ten weeks depending on the site and the chemical. Fly baits are also available, but they have a short life.

Chemicals may also be applied to where larvae are feeding, but contact with the larvae within composted and faecal material is often difficult to achieve.

Space sprays are often used in homes and are usually pyrethrins or synthetic pyrethroids in low pressure aerosol containers. Flies come into contact with the fine droplets of spray, causing a rapid reduction in the fly population. Food and food containers must be covered or protected during application.

MOSQUITOES

osquitoes are not only suckers of blood from humans and animals, but are responsible for the transmission of certain diseases, some of them fatal.

FEEDING HABITS & LIFE CYCLE

Mosquitoes have piercing and sucking mouthparts. They are attracted to their hosts once they detect warmth, moisture from the body and carbon dioxide being breathed out into the atmosphere. Sweat on human bodies may also be an attractant.

The female mosquito is the blood sucker. Irritation usually arises when the female injects saliva into the skin to serve as an anticoagulant to keep the blood flowing. The female requires a blood meal before it can lay eggs, which are placed on the surface of water. These hatch into

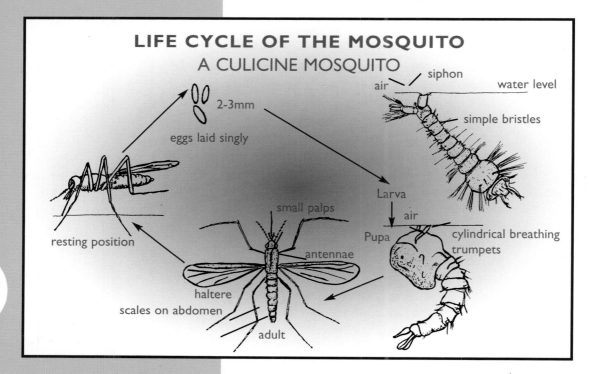

LIFE CYCLE OF THE MOSQUITO
A CULICINE MOSQUITO

2-3mm

eggs laid singly

air / siphon

water level

simple bristles

Larva

air

Pupa

cylindrical breathing trumpets

resting position

small palps

antennae

haltere

scales on abdomen

adult

aquatic larvae or 'wrigglers' which feed on minute organic particles in the water. When fully grown they pupate into mobile pupae known as 'tumblers'. The adult mosquito which emerges from the pupa has long legs and one pair of wings. The male mosquito is not a blood sucker, but feeds on plant secretions. The life cycle can be completed within one week, but is usually longer. During that time mosquitos can disperse several kilometres.

DISEASE

Millions of lives are lost throughout the world annually by the diseases transmitted by the female mosquito. The mosquito feeds on infected blood , and the microorganisms multiply within her body. These are then injected into a new host who has no infection.

There are several types of diseases transmitted by mosquitoes:

- protozoan diseases such as malaria
- worm diseases such as filariasis and dog heartworm
- virus diseases such as dengue fever, yellow fever, Murray Valley encephalitis and Ross River (epidemic polyarthritis) virus.

MOSQUITO SPECIES

The average homeowner cannot identify the many species of mosquitoes which occur throughout Australia. Some diseases are specific to certain species which occur in Australia. Of the common species and diseases:

- brown house mosquito (*Culex quinquefasciatus*) is a likely carrier of filariasis and dog heartworm
- common banded mosquito (*Culex annulirostris*) is a carrier of Murray Valley encephalitis, dog heartworm and Ross River fever
- domestic container mosquito (*Aedes notoscriptus*) is a carrier of myxomatosis and dog heartworm
- dengue mosquito (*A. aegypti*) is a carrier of dengue fever and dog heartworm
- common Australian anopheline (*Anopheles annulipes*) is a likely carrier of myxomatosis, malaria and filariasis.

31

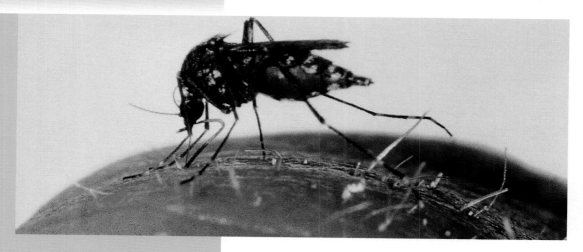

A female mosquito
obtaining a blood meal.

CONTROL

Non-chemical methods

Non-chemical methods can significantly reduce mosquito populations, mainly by the elimination of breeding sites, which can be anywhere water is held in pools, swamps, dams, rock pools, discarded car tyres and old tins. Guttering around a roof that holds water often for some weeks, is a major breeding area for mosquitoes. Door and window screens affect the entry of mosquitoes into houses, but do not reduce the numbers.

Chemical methods

Chemical control is used against the adults and larvae, but the treatment of larvae has a greater impact on the mosquito population. Treatment of larvae is often done by applying a thin film of oil on the surface of the water. The oil blocks the breathing tubes of the larvae and pupae.

Surface sprays of insecticides such as chlorpyrifos, deltamethrin, permethrin and propoxur, applied to surfaces on which mosquitoes rest, can be effective for up to a few months. Space sprays of non-residual insecticides can temporarily reduce adult populations inside homes.

Repellents such as diethyl toluamide prepared as lotions and aerosols give protection against mosquito bites for a few hours. Electric zappers are used mainly outside. Mosquito coils produce smokes which mainly repel mosquitoes. Some natural plant materials such as citronella have short-term repellent effects.

The elimination of breeding sites is the most effective long-term control.

BED BUGS

The bed bug (*Cimex lectularius*) has been a pest of humans for thousands of years, but since the 1940s when DDT was used in their control its numbers have decreased significantly. The improvement in hygiene in many countries has also contributed to the decline of the insect.

Although the adult and immature stages are blood suckers, bed bugs have not been linked with diseases of man. The developing nymphs and adults have piercing and sucking mouthparts often causing severe irritation as a result of the bites. Infestations have been more prevalent in hotels, motels and buildings where sleeping facilities exist.

FEATURES & LIFE CYCLE

Adult bed bugs are 4–5 mm long, have oval-shaped bodies and are rusty brown in colour. Prior to feeding

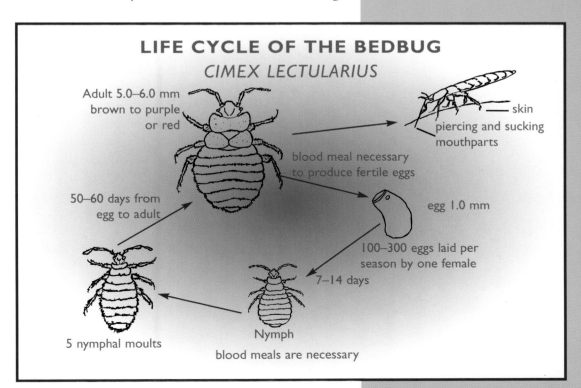

LIFE CYCLE OF THE BEDBUG
CIMEX LECTULARIUS

Adult 5.0–6.0 mm brown to purple or red

skin
piercing and sucking mouthparts

blood meal necessary to produce fertile eggs

50–60 days from egg to adult

egg 1.0 mm

100–300 eggs laid per season by one female

7–14 days

5 nymphal moults

Nymph

blood meals are necessary

33

An adult bed bug prior to a blood meal.

their bodies are rather flattened, but after a blood meal they are reddish in colour and are distended and oval in shape.

Before laying eggs the female must have a blood meal. Then, during egg-laying, she glues the eggs to surfaces near where she resides, each female laying over 200 eggs in her lifetime which may be up to 12 months. Once the young hatch they moult about five times through immature nymphal stages before reaching adulthood. A blood meal between each moult is essential to their continued development. The life cycle from egg to adult can be as short as 50 days, but this is dependent on the availability of hosts.

Males, females and the immature stages are all blood suckers, and usually obtain their host's blood a few hours before dawn. They then return to their hiding places in cracks and crevices in furniture near beds and in parts of mattresses. Discarded skins of developing nymphs and fine excreta are often found on sheets and when the infestation is heavy a bug-like odour may be present.

The bed bug usually injects an anticoagulant that can cause severe irritation in many victims. These irritations are often scratched leading to infection of wounds, but the bed bug does not usually convey diseases when it pierces the skin.

Bed bugs may be carried from sites of infestation in luggage, on clothing and in furniture. It is this way that infestations occur in hotels and motels.

CONTROL

While bed bugs are seldom found in houses where there is a high standard of personal and domestic hygiene, they can enter on clothing and in second-hand furniture. The adults and their eggs are then found in cracks and crevices. The eggs are glued on surfaces of walls and furniture. It is therefore important that when bed bugs are discovered that a thorough inspection be made prior to a chemical treatment.

Some of the chemicals used as sprays are organophosphates such as diazinon, fenthion and dichlorvos. Carbamates such a bendiocarb and propoxur also achieve control. Sprays of low-toxicity synthetic pyrethroids such as deltamethrin and permethrin also give very effective control. Sprays are applied to surfaces with special attention to the cracks and crevices in walls and furniture.

Where liquids cannot be used, dusts of bendiocarb or permethrin give good control, particularly when applied into wall cavities and furniture.

Where bed mattresses are infested they can be removed from the room and treated with a pyrethroid such as deltamethrin or permethrin. Where chemical treatment of a mattress is not acceptable for certain reasons the mattress can be completely wrapped in black plastic sheeting and placed on a frame off the ground in full sunlight for from three to four hours. A temperature of 50–60°C is usually reached which kills all insects present.

BED BUGS

WHAT TO DO

The following procedure should prove effective in the long term in most situations to achieve control of bed bugs:

1   Identify the pest to confirm it is bed bugs.

2   A person experienced in detection and control must conduct a thorough inspection of the area of infestation before proceeding with control procedures.

3   Treat the area with a careful application of insecticides of low toxicity to humans.

4   Ensure high standards of sanitation and hygiene follow as a prevention measure.

SPIDERS

Most homes are visited by spiders, but unfortunately their presence usually ends up in their death. Even when spiders are located in the garden their role in the control of pests of plants is not realised by many homeowners. Spiders play a very useful role in our environment by controlling other pests that they prey upon and should be regarded with respect.

Because the male funnelweb and female redback spiders are poisonous, most other spiders are regarded as toxic and aggressive, but this is not so for they are important components of the environment. Most other spiders are not aggressive.

FEATURES, LIFE CYCLE & HABITS

Spiders differ from insects structurally in that they have eight legs, their head and thorax is combined to form a 'cephalothorax', and there is a constriction between the cephalothorax and abdomen. They have a pair of 'palps' in the front which may be mistaken for a pair of legs, but they are sensory and in the male used during mating. The male is usually smaller than the female, has longer legs, a large terminal area on the palps, and a smaller abdomen. Spiders usually have three or four pairs of single-cell eyes: a feature which can often be used in their identification.

The female spider produces an egg sac in which the eggs are laid. The eggs inside the egg sac hatch so that when it is opened for inspection a large number of live spiderlings are present. Once the spiderlings emerge from the sac their growth involves several moults of their outside skin until they reach the adult stage. Most spiders

The male Sydney funnelweb is a poisonous ground-dwelling spider.

36

complete their development in one year but some ground dwelling species may take a few years.

Spiders are mostly nocturnal, leaving the security of their burrows or shelters to search for food at night. Webs of the web-spinning spiders are made at night to catch their prey. Spiders squeeze their prey with their fangs and the fluids are taken in through their tiny mouths. They feed on insects and other small arthropods, but cannibalism also occurs, particularly after mating when the female consumes the body fluids of the male.

Spiders produce silk or web from structures known as spinnerets on the rear end of their abdomen. This silk is particularly important to orb-weaving spiders which rely on their webs to trap insects. Ground-dwelling spiders lay silk on the ground near their burrows to trap insects and other small animals.

While the venom of some spiders such as the funnelweb and redback can cause the death of humans, most other spider bites cause only minor irritation and swelling. It is therefore essential that if a person has been bitten by a spider that an identification be obtained.

GROUND-DWELLING SPIDERS

Ground-dwelling spiders make holes in the ground and often live there with silken web material. Some of these spiders are regarded as poisonous, and in the case of the Sydney and northern rivers funnelweb spiders very poisonous.

Sydney funnelweb
Atrax robustus

This spider is often encountered in most suburbs of Sydney, and along coastal New South Wales. It is black with fine reddish hairs, and favours dark places to make its tunnel. The male's body is about 20 mm long: the female's larger, at 30 mm. It has a long life cycle of three to five years, with some females living even longer. The male is identified by having a spur on the second front pair of legs. The male is more toxic than the female, and more active in late summer and autumn when it is seeking a mate.

The female Sydney funnelweb is also poisonous, and larger than the male.

37

Sydney funnelweb female defending the entrance to her home in the ground.

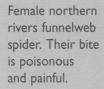

Female northern rivers funnelweb spider. Their bite is poisonous and painful.

38

Northern rivers funnelweb spider
Hadronyche formidabilis

This is another poisonous species which occurs along the north coast of New South Wales, and is found in holes in trees. They are black with reddish hairs, the male measuring 35 mm, and the female 50 mm in length. They have a life cycle of three to five years.

The egg sacs of the Sydney funnelweb spider contain many young.

39

Sydney brown trapdoor spider
Misgolas rapax

This is the most often encountered trapdoor spider around Sydney, although the hole in the ground in which it lives may not have the 'trapdoor' cover like some other species of trapdoor spiders. The Sydney brown trapdoor

A male Sydney brown trapdoor spider.

A female Sydney brown trapdoor spider. The bite is non-toxic but may produce some pain.

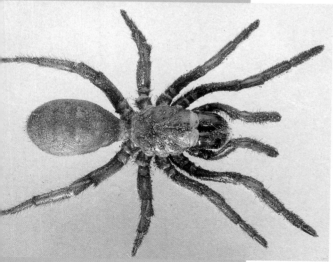

is not considered poisonous to humans. Both male and female are brown to dark brown and covered with hairs. The male can be identified by its boxing glove-like palp, and two spines on the inner surface of the front legs. The male is usually 20 mm in length: the female 25 mm.

Mouse spiders
Missulena spp.

Mouse spiders occur over most of Australia. They are not aggressive, and the female lives in a hole in the ground, often with a door over the entrance. Both sexes are black but in one species the head area of the male is a bright red with black body. The males have longer legs than the female.

The eastern mouse spider (*M. bradleyi*) is often found in swimming pools and may be mistaken for the Sydney funnelweb. The male measures 15 mm while the female is 20 mm in length.

In this species of mouse spider the male's cephalothorax is red.

Female mouse spider. The bite is painful and also toxic.

Wolf spiders
Lycosa spp.

Wolf spiders are frequently encountered over most of New South Wales and other states. They live in holes in the ground and are mottled grey colour, the cephalothorax being referred to as 'Union Jack' in appearance. As in other spiders the male has longer legs than the female. Their bites can be painful and are regarded as toxic. The male measures 15 mm and the female 20 mm in length.

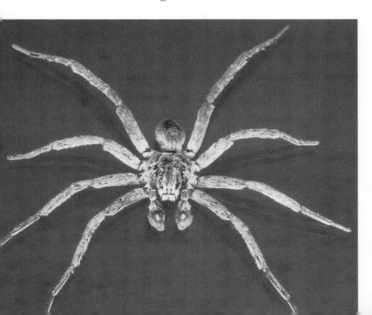

The female wolf spider is larger than the male. The bite may cause some pain.

41

The male wolf spider has large palps in front.

A female garden orbweaving spider. They rarely bite and are not poisonous to people.

A female St. Andrew's cross spider with its legs placed in pairs like a cross.

A female leafcurling spider secure in a leaf.

ORBWEAVING SPIDERS

The spiders which construct webs produce two types of silk. One type is for the main part of the web structure and is intended to catch or snare flying insects. It is of a sticky or adhesive material. The other type forms the guy ropes which are attached to structures or vegetation, and it is of a non-adhesive material. An 'orb' web consists of threads radiating from a central point, supporting a spiral of sticky web to catch flying insects. After one night both the central sticky area and the dry silk need replacing. After squeezing the body fluids from their catch the spider wraps the body in silk and leaves it hanging in the web area. All the orbweaving spiders are non-poisonous to humans, and not aggressive. They have only one generation a year.

Garden orbweaving spiders
Eriophora spp.

This spider is dark to light brown, and hairy. It hides in foliage during the day and constructs its web at night, usually in shrubs and trees and across pathways. It is not aggressive, it seldom bites and it is non-toxic. Males are 10 mm in length and females 25 mm.

St Andrew's cross spider
Argiope keyserlingii

This spider has a brown cephalothorax with a yellow and brown striped abdomen. It hangs in its web with legs in the shape of a cross. It is not aggressive and is non-toxic. Usually the males are 5 mm in size: the female 12 mm.

The two-spined spider (female shown here) occurs outside and is non-toxic.

Leafcurling spider
Phonognatha graeffei

This is a brown spider with yellow markings, which resides in a curled leaf or paper at the centre of its orb web. It likes to suspend its snares along paths and insect trails in open forest. Another favoured spot is near paling fences in suburban gardens. It has a small cephalothorax and large abdomen. The male is 4 mm and the female is 7 mm in length.

Two-spined spider
Poecilopachys bispinosa

This spider has a brown and cream body with reddish legs. It makes an egg sac larger than its own body, and has two dorsal spines on its abdomen. It can be found hanging in its web among young trees and shrubs within a couple of metres from the ground often in heath and woodlands. It is a totally harmless spider. The male is 3 mm and female 7 mm long.

Golden orbweavers
Nephila spp.

This orbweaving spider has a yellow and purple body which is rather velvety with fine hairs. It hangs in

Male and female (larger) golden orbweaver spiders.

strong strands of web which often have a golden sheen in the sunlight. It is not aggressive and non-toxic, the male being 5 mm and female 30 mm in length.

OTHER SPIDERS

Redback spider
Latrodectus hasselti

The female redback spider is extremely poisonous. It has a black velvety body with a red stripe on the upper surface of the abdomen. It makes loose, untidy webs among leaves, rubbish, under houses, in tins, tyres, stacked articles and outside toilets. The male spider is very small and harmless and measures 3 mm in length. The female can be up to 12 mm in length.

Black house spider
Badumna insignis

This is a poisonous spider whose bite can cause pain, nausea and sweating but is not lethal. It is dark brown in colour and its natural habitat is in the bark of trees. They help to control flies and mosquitoes around the house, but their webs are unsightly. They like to sit in the centre of their webs which are often made in

Bites from the female redback spider are very poisonous and painful. The male is smaller and harmless.

A female black house spider. Their bites cause pain and discomfort.

45

window sills, under guttering and around toilets. The male is usually 8 mm and female 12–18 mm in size.

Whitetailed spider
Lampona cylindrata

This is another poisonous spider whose bites cause pain and have been associated with blistering and necrosis (where the cells on and just below the surface die and become scaly). It is black with a white patch at the end of its abdomen. The male is about 12 mm while the female can be up to 20 mm in length.

This spider is a wanderer and a hunter. Its main diet is other spiders, including the black house spider. They are found under bark and logs in the bush. When it lives in houses, it searches for its prey in the early evening and shelters during the day in bathrooms, the tops of walls and in furniture.

Huntsman spider
Isopoda immanis

The huntsman spider is not aggressive, but if provoked its bite can be painful although no symptoms of poisoning follow. Its front two pairs of legs are larger than the rear two pairs, and its body has a rather flattened appearance. Its colouring can vary from browns to greys and buff, and it is often mottled. The huntsman can move sideways. The male is usually 25 mm in size and the female 35 mm.

They normally live under bark of trees during the daytime and emerge at night. They often enter houses where they are seen on walls. They are a useful spider in that they feed on insects.

Netcasting spiders
Deinopis spp.

These are non-toxic spiders which hold their nets between their front legs for catching insects. They resemble small sticks, and have long legs. They are a brown, often mottled colour. The male measures 8 mm and the female 15 mm, but the sizes are variable with different species.

A female netcasting spider. It catches its prey in the folded net within its legs.

46

A female whitetailed spider. Bites are painful and sometimes cause localised paralysis.

A female huntsman spider with its egg sac. Huntsmen bites are very rare, and not poisonous.

47

Jumping spiders (female shown) are often encountered around houses, but are non-poisonous.

Jumping spiders
Myrmarachne spp.

Jumping spiders are non-poisonous and are the only spiders that can actually jump from a flat surface. They usually seek food during the day. Their front legs are very large and strong. The male is usually 6 mm and the female 8 mm in length.

Spider anglers
Ordgarius spp.

These spiders are non-toxic and make their webs on leaves. They produce a thread with sticky globules to catch insects. They are cream in colour and the male measures 3 mm, the female 12 mm.

CONTROL

1   Identify all spiders found in the house. Kill only those that are regarded as poisonous to humans.

2   Take care when using insecticides in gardens, as this can cause funnelweb spiders to enter houses. Wet weather may also cause ground dwelling spiders to enter houses. Ground-dwelling spiders are also often found in swimming pools and can be alive even after hours at the bottom. Do not remove with the hands.

3   Carefully inspect all footwear and clothing left outside before dressing, particularly in summer and autumn.

4   Inspect all toys, clothing, footwear and other articles left outside at night before bringing them into the house.

5   Use a vacuum cleaner to remove any webs from under eaves and around windows. Release harmless spiders into a more suitable environment.

TICKS & MITES

There are many species of ticks, but the species which are most often encountered along the east coast of Australia are the Australian paralysis tick (*Ixodes holocyclus*) and the brown dog tick (*Rhipicephalus sanguineus*). Both ticks can occur on many different animals, as well as on humans. The Australian paralysis tick causes the death of animals, particularly dogs and cats, and illness in humans. Whenever illness of humans and animals is suspected, the cases should be referred to the relevant medical or veterinary surgery.

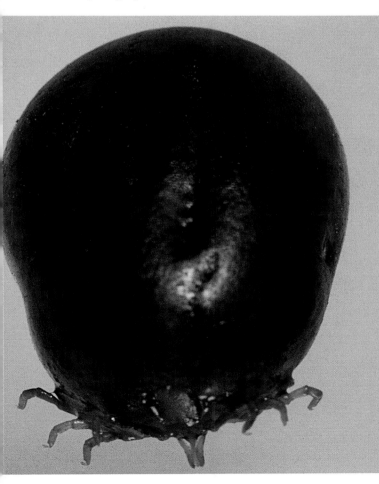

This female paralysis tick (left) has recently had a blood meal. The male tick (above) does not feed on blood and is not a pest of humans or animals.

A female paralysis tick embedded in skin and feeding on blood.

AUSTRALIAN PARALYSIS TICK

The Australian paralysis tick has four developmental stages: egg, larva, nymph and adult, the last three being as external parasites of man and animals. Only the female tick is a blood feeder. She inserts her mouthparts and injects an anticoagulant so that the blood will continue to flow. It is the anticoagulant which is apparently toxic. The mouthparts are barbed to hold the tick in the skin.

The symptoms of paralysis tick lodgment in the skin can include:

- pain and irritation at the site of attachment
- loss of appetite
- lassitude (sleepiness) and depression
- discharge from the eyes.

Once located the tick should be removed. Because of the flatness of the mouthparts and the backwards projecting barbs, it is necessary to use forceps. Grasp the mouthparts on the head, and move them sideways before withdrawing the tick. Do not squeeze the body or apply any chemicals, as more toxin will be injected into the bloodstream.

TICKS
WHAT TO DO

The elimination of paralysis ticks from an area is not possible without the use of insecticides and the removal of natural host animals. However, the following procedures will reduce the numbers of ticks:

1 Reduce or eliminate moisture from known tick areas.

2 Reduce weeds and unwanted low vegetation. Keep grass cut where possible.

3 Exclude or discourage (but do not kill) bandicoots, as they are natural hosts of the paralysis tick.

4 Spray areas known to harbour ticks, using an approved insecticide such as bendiocarb, carbaryl, chlorpyrifos or diazinon. The label will show whether it is approved for this use and also application directions.

5 Inspect pets daily (or every second day). When ticks are located they should be carefully removed with forceps.

6 Wash animals in a solution containing an insecticide approved for that purpose (for instance. 'Frontline'). Follow all the directions on the label of this treatment.

MITES

Mites are tiny creatures about 1–2 mm long which favour moist conditions. For this reason they are seldom encountered in air-conditioned buildings. Mites occur on food, particularly in moist situations, but more commonly it is the mites of birds and other animals which cause temporary irritation of humans. Those mites which occur on food have a type of chewing mouthpart, while those which occur on animals and plants have piercing and sucking mouthparts.

Starling mites

Are found on caged and aviary birds, as well as in nests in roof cavities. When birds leave their nests the mites also leave to seek other hosts. They then often become temporary parasites of humans.

51

Poultry mites

Poultry mites can also be carried from poultry and cause irritation to humans.

Bryobia mites

These are found on vegetation, and can also cause temporary irritation.

Straw itch mite

This mite occurs in straw and straw mattresses, and is also found on certain insects such as borers, and is a temporary pest of humans.

Grass itch mites

These may be found in reddish clusters in the ears of cats. They can cause severe irritation in humans which may last some days, often requiring medical attention.

The scabies mite

This mite tunnels in the upper layers of skin causing intense irritation. It usually occurs where there is poor hygiene, for it is conveyed by human contact and in clothing. Medical attention is needed for the control of this mite as well as strict personal hygiene.

The house dust mite
Dermatophagoides spp.

These are present in most houses and those who have allergic responses such as asthma and hay fever can be constantly affected by their presence.

MITES
WHAT TO DO

Where mites are located in a bird's nest in a roof or similar situations a carbamate or organophosphorus insecticide will prove effective. Check that the label data includes mites in its nominated range of pests.

Chemical control of dust mites is mostly unreliable, and the householder should vacuum thoroughly and repeatedly. Dry-cleaning and laundering of clothing and bedding also complements vacuuming. Contents of the vacuum cleaner bag should be heat treated in the sun in black plastic after each use and not left in the bag.

CHAPTER 9

LICE

Head, body and crab lice are widely distributed throughout the world. When they infest humans the condition is known as 'pediculosis'. In Australia the head louse and crab louse are of common occurrence, but the body louse is seldom encountered.

Head louse
Pediculus capitis

This louse is about 2.0–3.0 mm long. It has three pairs of legs with fine claws on the ends which facilitate their movement through hairs on the head. It feeds mainly at the back of the head and behind the ears. It is in these places that the adult lice lay their eggs, sometimes called nits, attached to the strands of hair. The eggs usually hatch in five to ten days. Head lice are spread during personal contact and sharing combs, brushes, hats and scarves and pillows. They are often encountered in schools where children from homes of varying standards of hygiene have contact.

Body louse
Pediculus humanus

These are almost identical in appearance to head lice, but they differ in their habits and feeding locations in humans. While body lice occur in many parts of the world, they are not pests in Australia. Body lice transmit epidemic typhus, trench fever and relapsing fever in much the same way that the oriental rat fleas transmit bubonic plague. (Historically the body louse has significantly affected the outcome of certain wars where it has caused illness and deaths due to typhus and trench fever.) Epidemic typhus occurs when the wound is scratched and the typhus organism (*Richettsia* sp.) is conveyed from the louse faeces or the crushed louse body to the wound.

53

Crab louse
Pthirus pubis

This louse is about 1.2–2.0 mm long, and has a wide body. It has large claws on its middle and hind legs, and resembles a minute crab. It mostly infests the pubic and perianal regions where the hairs are more widely spaced than on the head. They spread mainly by sexual contact, but also from toilets and bedding. They are not associated with the transmission of diseases.

LIFE CYCLE OF THE LOUSE

THE PUBLIC LOUSE
PHTHIRUS PUBLIS

THE BODY LOUSE
PEDICULUS HUMANUS

eggs 0.8mm attached to clothing

6-10 days to hatch

adult 3-4mm

10 days

Nymph appears very like the parent in clothing on the body

LICE
WHAT TO DO

When suspected louse infestations are detected it is important that the cause be identified, usually by the medical profession. Treatment usually involves lotions, washes and aerosols — whichever is appropriate for the particular infestation. The materials usually prescribed or recommended contain pyrethrins, bioallethrin, carbaryl or maldison. Once treatment has achieved control, high standards of personal hygiene, particularly in schools, should reduce the chances of re-infestation.

CHAPTER 10

FOOD PESTS

Pests of food commence causing problems at the sites of production on farms, and there is continual exposure through silos, mills, food processing plants, storage facilities, bakeries, shops, restaurants and homes. Special protection and control is usually required at each site.

After the purchase of various stored foods, pests can be brought into the home in infested products. Many foods used in the home have their origin in various grains such as maize, wheat, oats and rice and these may be attacked by many different food pests. As many pests of grain such as rice pass most of their life cycles within individual grains, their presence is often not detected until the adults emerge.

The most often encountered pests of stored products are beetles and moths and their larvae. Some of the insects which infest food are the drugstore beetle, cigarette beetle, lesser grain borer, rice weevil, granary weevil, hide beetle, saw-toothed grain beetle, yellow mealworm, angoumois grain moth, Mediterranean flour moth and Indian meal moth. When seeking more information on the habits and control of stored food pests, the identification of the insect is an important first step. This often leads to the source of the infestation which may be in a different food in a different location.

DETECTING INFESTATION

The following characteristics betray the presence of insects in food:

- faecal material, usually as a fine dust
- food webbed together by fine strands of silk. This indicates the larvae of moths are present.
- visible beetles, moths or their larvae
- in the case of grain foods such as rice, whole but hollow grains
- an odour in the food that is heavily infested.

The saw-toothed grain beetle has spines on the edges of its thorax. It is found in grain products

The adult lesser grain borer has its head concealed beneath its thorax. Its larvae attack and live inside grain.

Rice weevils are pests of grain.

MINIMISING INFESTATIONS

Infestations of food in homes can be prevented, unless the food purchased is already infested. Hygiene in kitchens and other storage areas complements the following methods of reducing infestation by stored product pests.

1. Carefully treat all small food parcels brought into the house in a microwave, programmed so that the food is not adversely affected.

2. Transfer food to sealable kitchen containers for storage, after heating if required.

3. Mark the storage dates on food containers and use the older ones first.

4. Seal food containers immediately after use.

5. Frequently inspect food in containers.

Chemical treatment of food in eating facilities is not recommended. Where food particles occur in kitchen cupboards these can be removed using a vacuum cleaner. Where insects exist in cracks and crevices in storage cupboards pyrethrin sprays of low toxicity are effective, but stored food must be removed or containers must be sealed.

FOOD MOTHS
WHAT TO DO

The infestation of stored food, particularly grain products, is often due to the larvae of moths. These are usually detected by the presence of web-like material which bonds together the food and the faecal material of the larvae. When such infestation occurs, the food should be discarded. Larvae are able to penetrate some thin plastic containers and packaging by chewing through the surface. The life cycle of a moth is usually three months and shorter during summer. Consequently, infestation can occur several times in one season. To control food moth larvae that have infested kitchen cupboards, pyrethrin sprays may be applied.

CHAPTER 11 · CARPET BEETLES

Carpet beetles are common pests of articles of animal origin in homes. The larvae feed on carpets, underfelt, fur products, wool, silk and hair. They will also feed on dried meat products, cereals and dried insects. The black carpet beetle may even be found in roof cavities where rodents and possums have died.

Usually there is only one generation a year. The eggs are laid near a food source. The young then pass through larval and pupal stages, which may take 9–12 months, before becoming adult beetles. The beetle then leaves the larval site to feed on pollen and nectar of flowers. It is then that the beetles enter houses on flowers or fly inside to lay eggs near a food source for their larvae.

The two main carpet beetles are the variegated carpet beetle, which is about 2–3 mm long, and the black carpet beetle, which is 3–5 mm long. Most household insect pests can not tolerate dry conditions: these beetles can, even the less humid conditions of inland Australia.

Variegated carpet beetle is a pest of furs and woollen products such as felts and carpets.

The variegated carpet beetle is a mottled colour of yellow and black, and the shape of the body is similar to that of a ladybird beetle. The larvae are 4–5 mm long when fully grown. They are broadest at the rear end and are covered with stiff brown bristles.

The black carpet beetle is shiny black in colour and rather elongated. Its larvae are 7 mm long when fully grown, reddish brown and covered with bristles. Its body tapers towards the end and has a clump of long hair from the tip of the abdomen.

57

CONTROL

Non-chemical control

Prior to implementing any control measures, an inspection of all accessible areas of a house is essential, as beetles can be present in roof cavities in the bodies of rodents or possums. They also eat other insect bodies which often occur in roof cavities. The edges of carpets and behind furniture often conceal infestations, for they are not accessible during normal vacuuming.

Cut flowers to be used in houses should be examined for carpet beetles, as this is a significant cause of their distribution. Soiled clothing, even if of synthetic fibres, should be washed before storage. When infestations of clothing are located the article may be wrapped in black plastic and placed in full sunlight for two to four hours. The temperature achieved inside kills all stages of carpet beetles, but does not damage the fabrics.

Chemical control

A thorough inspection and vacuuming of areas of a house, including under and behind furniture, should be done prior to the application of surface sprays, aerosols or dusts. Before applying any chemicals they

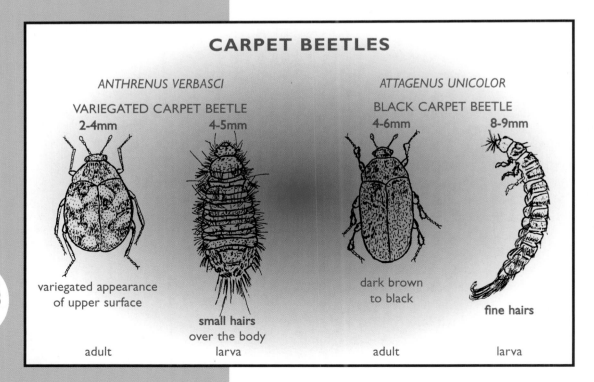

CARPET BEETLES

ANTHRENUS VERBASCI

VARIEGATED CARPET BEETLE
2-4mm   4-5mm

variegated appearance
of upper surface

small hairs
over the body

adult   larva

ATTAGENUS UNICOLOR

BLACK CARPET BEETLE
4-6mm   8-9mm

dark brown
to black

fine hairs

adult   larva

should be tested on a small area of the fabric or other type of surface to determine if staining or other problems may occur.

Several insecticides are registered for the application to areas infested by carpet beetles. Specific uses appear on the labels. Organophosphorus insecticides such as diazinon and dichlorvos and the carbamates bendiocarb and propoxur may be used. Also the synthetic pyrethroids deltamethrin and permethrin may be used as sprays. Aerosols of permethrin and pyrethrins are also approved in most states. Dusts can also be used, but these must be carefully brushed into the surface of carpets so that they penetrate to where the larvae are feeding.

Wettable powders are often used on carpets to avoid stains, and also because the particles can be vacuumed up at a later date. When activity is detected in roof cavities bird nests, rodent bodies and other breeding material should be removed before using an insecticidal dust or sprays.

CARPET BEETLES
WHAT TO DO

When an infestation of carpet beetles is detected the following sequence of steps is recommended for control:

1   Carry out a complete inspection of the house, including the roof cavity, because carpet beetles spread quickly from one area to another.

2   Thoroughly vacuum all carpets and other sites of infestation. It is best to move furniture to allow special attention to be given to the edges of carpets.

3   Treat the contents of the vacuum cleaner with an insecticidal aerosol prior to it being emptied, as it may contain carpet beetle larvae and eggs depending on the time of year. Do not return the vacuum cleaner bags and untreated contents to the interior of the house.

4   The services of a pest control technician may now be obtained to treat the areas of infestation. Minor infestations can also be controlled by the householder.

Once controlled, regular inspections and vacuuming of carpets, and careful inspection of flowers and fabrics should keep the home free of carpet beetles.

CHAPTER 12

PESTS OF FABRIC

The pests of fabric which are usually encountered in the home are clothes moths, carpet beetles, silverfish and book lice. Because carpet beetles are a serious pest of floor covers and many other articles of animal origin they have been considered separately, in the preceding chapter.

Skin and hide beetles are sometimes found in furs and also in roof cavities in possum and rodent bodies (from left: adult, larva and pupa).

CLOTHES MOTHS

Not all moths are pests of fabrics. The casemaking clothes moth (*Tinea pellionella*) and the webbing clothes moth (*Tineola bisselliella*) attack materials mainly of animal origin such as woollens, felts and furs. These two moths generally occur in the more humid coastal regions, where their larvae can do extensive damage to carpets and clothing.

The adult clothes moths are small, usually up to 10 mm long and often yellow or gold to buff colour. They have narrow wings that are fringed, and siphon-type sucking mouthparts, although adults do not feed. The larvae which damage the fabrics have chewing mouthparts, and are white in colour with a dark head and six legs. The larvae of the both species are caterpillar-like,

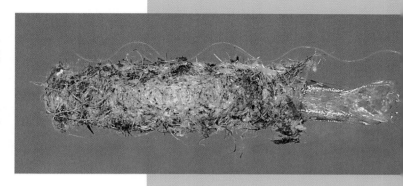

Casemaking clothes moth cases are often found on walls and behind furniture.

The larva of the casemaking clothes moth on fabric.

61

but the casemaking clothes moth larvae live in cases made of silk and fibres from material in which they are feeding. The colour of the case may give some indication of the material it is feeding on, especially if it has migrated away from the feeding site. The webbing clothes moth moves freely within its food material, without a protective case.

Female clothes moths usually lay their eggs on materials that will provide a suitable food source for larval development. Eggs hatch in five to ten days, and then the caterpillar-like larvae feed in dark undisturbed places. The larval or feeding period is usually about two to six months. The larvae of the webbing clothes moth pupate within the material in which they feed while the casemaking clothes moth larvae leave the food material in their cases and pupate behind cupboards, on walls and picture rails.

CONTROL

Clothing affected by clothes moths can be sterilised, by washing or wrapping in black plastic and exposing to the sun for two to three hours.

Chemical control using sprays of deltamethrin, permethrin and propoxur are effective when applied to clothing storage areas. When clothes moths attack floor coverings it is important to give special attention to the edges and behind furniture for these are favoured sites for breeding, as very little disturbance occurs.

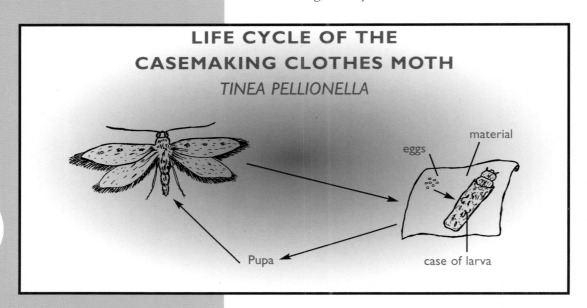

LIFE CYCLE OF THE
CASEMAKING CLOTHES MOTH
TINEA PELLIONELLA

material

eggs

Pupa

case of larva

SILVERFISH

Silverfish are primitive insects which belong to the insect Order Thysanura. Several species enter houses and infest areas where starchy materials such as paper, books and fabric are stored. They also occur in roof cavities. They are often found fallen into baths and sinks, for they cannot climb smooth surfaces.

They are covered with small scales and are somewhat flattened. Silverfish may live for up to four years. All stages moult, including the adults.

CONTROL

Inspecting goods brought into the home (such as second-hand books) may help to avoid infestation. Books, files and papers should be stored in light and airy conditions where possible. Foods kept in kitchen areas should be kept in sealed containers and crumbs and scraps cleaned up thus reducing the suitability of that environment to silverfish.

Sprays of chlorpyrifos, cypermethrin, deltamethrin and permethrin are effective in the control of silverfish. Once controlled the inspection of various articles such as books and files should keep an environment free of silverfish.

Silverfish are pests of starchy materials, such as wall paper and book bindings, and often found in ceilings.

63

BOOKLICE

These are also called 'psocids'. They are very small insects about 1–2 mm long and feed on moulds which occur in various foods, particularly where the humidity is high. Because of their size they are difficult to detect unless present in large numbers.

CONTROL

The reduction in humidity by improved ventilation is the most effective control in the long term. Where there is air conditioning, the atmosphere is comparatively dry and mould does not grow to support book lice.

While many surface sprays such a deltamethrin and permethrin are effective there are no insecticides specifically registered for booklice.

LIFE CYCLE OF THE BOOK LOUSE
PSOCID

mandibulate mouth parts

Adult 1-2mm

egg slightly flattened

moults

Nymph
found on wood, bark, fungal growths
feed on mould or paper

64

CHAPTER WASPS

The European (*Vespula germanica*)wasp was introduced into Australia apparently during the latter half of the twentieth century and is now a pest in most Australian states. It also occurs in New Zealand where it attacks weak hives of honey bees affecting honey production. This is also expected to occur in some parts of Australia as it spreads northwards and westwards. Unlike honey bees, European wasps can sting many times when disturbed, particularly if this occurs in or near their nest. Their stings, particularly those from a whole swarm, can require medical attention.

The English wasp (*Vespula vulgaris*) has been present only in Victoria for some years and is very similar in appearance to the European wasp. It is therefore important to have wasps identified before control measures are applied.

The European wasp is an introduced pest which can inflict multiple painful stings.

EUROPEAN WASP

LIFE CYCLE AND HABITS

The workers of the European wasp in the colony are sterile females and measure 12–15 mm, with full wings which cover their bodies. They have black bodies with horizontal yellow bands. The queens and males are larger than the workers, being about 20 mm long.

Recognition of the nest is important. They are made in the ground often near an old tree stump, in rockeries and landscape garden material. They have also made nests in roof cavities and walls from which they continuously leave to collect food.

To obtain their two basic food needs of protein and carbohydrate, the wasps seek food from flowers and fruit, and also prey on many insects such as flies and honey bees. A single female wasp establishes her colony in a selected location and soon commences to lay eggs.

In a developing colony the larvae, which are fed by the queen, pupate and then emerge as adults which take over the duties of the colony, including the feeding of the young. At the end of a summer a nest can contain over 10 000 individuals. The nests can vary in size from 15 cm to over 4 metres. At the height of their development they may contain over 4 million cells and over 100 000 workers.

CONTROL

The services of a pest control technician are recommended as being the safest and most effective means for the control of European wasp.

Control measures are best carried out at night, because the wasps remain in the nest. The nest can be covered with plastic sheeting and the insecticide introduced. In this way the wasps cannot leave the colony reducing the possibility of the operator being stung. However, it is important that body and facial protection is adopted when controlling the wasps.

Isolate and destroy the nest using a pesticide registered for this use. Propoxur and carbaryl dusts are effective in controlling the wasps in their nest. An aerosol mixture of propoxur and dichlorvos is also effective. The nest can then be removed and destroyed in case any wasps have survived the treatment.

PAPERNEST WASP

Papernest wasps are social insects with female, male and worker castes living in a papery nest, which is often found under the eaves of a house attached to the surface by a small strong stalk. The adults prey on other insects, but if disturbed can sting several times.

If the nests of paper nest wasps are to be removed it is safest to do so at night. Cover the nest with plastic, and then cut the stalk attachment.

PESTICIDES & PETS

During the last 25 years, most of the more toxic pesticides used to control household pests have been phased out in Australia. These included the organochlorines, some organophosphates and poisonous inorganic chemicals. Nevertheless, whenever any pesticides are used, the precautions written on the labels of containers must be followed.

Insecticides, such as most of the pyrethroids (permethrin, deltamethrin), have a good safety record. In recent years, insect growth regulators such as triflum-uron, which affect the developing stages of certain insects, have been used for some household pests, but these are only mildly toxic for warm-blooded creatures.

The most often encountered poisoning cases treated by veterinary surgeons are the result of anticoagulant rodenticides (rodent baits) such as brodifacoum and bromadiolone which cause internal haemorrhage and death by preventing blood clotting. Anticoagulant rodenticides must be stored away from children and pets. It is important when setting rodent baits that they are placed in containers which are accessible to rodents but not to children or pet animals. When using insect pesticides the following precautions are strongly recommended:

1 All pet animals, including birds in cages, should be removed from the house where the pesticide is being applied.

2 Animals should be allowed to re-enter the treated area only after pesticide has dried.

3 Where soil barrier treatments have been used to control termites, animals should not be allowed to dig or rest in treated soil.

4 Label instructions must be read and followed.

When using rodenticides, the following precautions should be taken:

1 Place baits in sites accessible to rodents, but not pets. Roof voids are often the safest.

2 Place rodenticides in a container with small access for rodents but not for pets.

3 Dead rodents or rodents affected by the anticoagulant poison may be eaten by pets and thus cause secondary poisoning. Such rodent bodies should be collected and disposed of safely. If a pet animal has consumed any anticoagulant rodent bait, it must be referred to a veterinary surgeon as soon as possible. Most of these poisoning cases are treated with vitamin K1.

RATS & MICE

THE ROOF RAT
RATTUS RATTUS

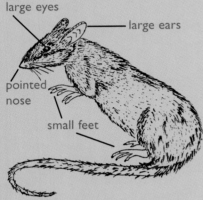

large eyes

large ears

pointed nose

small feet

tail longer than the head and body

THE NORWAY RAT
RATTUS NORVEGICUS

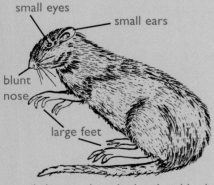

small eyes

small ears

blunt nose

large feet

tail shorter than the head and body

Rats and mice are mammals of the Order Rodentia (or rodents). There are three rodent pests of houses in Australia, all introduced species. They are the Norway rat (*Rattus norvegicus*), the roof rat (*Rattus rattus*) and the house mouse (*Mus musculus*).

Rats and mice require food and shelter like most animals, and these are provided in buildings, particularly during late autumn and winter when they enter houses. These conditions also complement their breeding. Once inside buildings they often make their nests in walls and roof cavities. The nests are usually composed of paper and other soft materials such as insulation batts. In the case of the rats, particularly the Norway rat, burrows are made in soil often near buildings and garbage disposal areas. Rats and mice are also very good climbers, being able to ascend rough walls, pipes, trees and vines, and walk across cables from one structure to another.

Rats and mice are mostly active at night and eat a wide range of foods. The senses of rodents such as smell, taste, hearing and touch are very keen, but their sight is poor and as a result they tend to remain close to various surfaces. They are sensitive to all environments, and move around the edges of rooms rather than across them. Rodents, especially rats, have a fear of new objects (neophobia) in their search for food and usually use their same tracks. This is important when baiting and trapping for it may be some days before they investigate a bait station or a trap. Mice are likely to investigate more quickly any changes in their environment compared to the two species of rats.

Grooming by licking their fur and feet is a normal practice for rodents. The action of the tracking powders containing anticoagulants is dependent on this.

PEST STATUS

Because of the diseases they carry, rats and mice have contaminated food which is then destroyed. This occurs at the site of harvesting, storage and food manufacture, as well as in homes, restaurants, hotels and motels. The diseases are usually the result of faecal, urine and fur contamination.

Rats and mice also cause physical damage. To keep their front teeth to a functional size rodents must gnaw, and they do so on many products including timbers, furniture and various articles in the house. Electrical wiring has been damaged in this way, resulting in house fires.

SPECIES

Norway rat *Rattus norvegicus*

This is the largest of the three, and is a major pest in most human environments where food and shelter are available. It is often present in food handling facilities, sewers, garbage areas and on most types of farms. Unlike mice and roof rats, Norway rats make burrows in the soil and have concealed escape holes known as 'bolt holes'. In cold weather, particularly in highland areas, they will live inside buildings making their nests in walls and roof cavities.

The Norway rat weighs about 450 grams. It has a blunt nose area, its tail is shorter than its body length, and its ears are close-set and small. Its fur is a coarse brown colour. It lives about one year and during that time produces five to six litters each having about eight young. When other rodents are also present the Norway rat is the physically dominant species.

Roof rat *Rattus rattus*

The roof rat is usually found in city and suburban areas where the human population density is higher than that in country. They usually nest indoors, but also outside where the vegetation gives them nesting security. When in buildings they nest in roof and wall cavities. They are good climbers and travel up and down pipes and across various forms of insulated wiring. Roof rats prefer vegetables, fruit and cereals.

The roof rat weighs about 250 grams, has a pointed nose, and large ears which are almost hairless. It is a uniform colour of grey, black or brown fur and

RODENTS AND DISEASE

In the fourteenth century the roof rat contributed to the deaths of over 25 million people in Europe by being a carrier of the bubonic plague (Black Death) which was passed on to humans by the oriental rat flea. The black death bacterium *Yersinia pestis* is still present in the world, but now more is known about its behaviour.

Rats and mice can transmit the following diseases to humans:

● Weil's disease from rodent urine and faeces

● murine typhus fever, conveyed from rats to humans via fleas

● Salmonella food poisoning

● favus, a fungal skin disease passed from mice to dogs and cats, and then to humans

● rat-bite fever, a bacterial infection, from any rodent bite

● trichinosis worms, passed on from poorly cooked pig products from pigs which have eaten infested rodents.

Roof rats helped spread
the Black Death.

The tail of the roof rat is
longer than its body.

may be white underneath. Its tail is longer than its entire body. The life expectancy of the roof rat is about one year during which time four or five litters are produced each having six to eight young.

House mouse
Mus musculus

The house mouse is also known as the 'field mouse'. When it is located outside it is a yellow-brown colour, paler than those living inside which are usually a darkish grey colour, with lighter grey on the belly. As it is a small rodent it obtains easy access to the home and nests in walls, cupboards and roof cavities. They make small burrows when outdoors. They have a low water requirement and feed on many foods such as grains, fruit and various animal and human foods. They are mainly nocturnal, but will feed during the day and will be attracted to new foods rather quickly. Contamination of food and kitchen utensils with their urine and faeces occurs whenever mice are present. Mice can reach plague numbers on farms and properties in rural areas where they feed on stored products of grain and vegetables.

The house mouse weighs about 20 grams, lives for about one year during which time it has about six to ten litters, each having five or six young. It has rather large, hairy ears, and a pointed nose, and its tail is about the length of its body.

DETECTION

The presence of rodents is often detected from their damage, odour and faecal droppings. Droppings about 18 mm indicates the Norway rat, 12 mm for the roof rat and 3–4 mm for the house mouse. The following signs should always be investigated to confirm rat and mouse presence prior to trapping or baiting:

● runways, showing up as greasy nooks on furniture and walls from the rodents' fur

● urine stains on surfaces of floors and cupboards

● disappearance of food

● sounds, often occurring at night, which may include squeaking and fighting

● nests behind cupboards and lounges made of paper and rags, and sometimes snail shells.

Pets are sensitive to other animal intruders and often bark when rodents enter houses or are active.

Once the rodent activity has been detected it is then important to determine which rodent is present, for this will determine the control.

CONTROL

Non-chemical control

The prevention or termination of rat and mouse infestations by sanitation and hygiene is the most satisfactory control procedure. The use of vacuum cleaners to reduce food particles on floors is important, but must be accompanied by using sealed food containers and also tight-fitting lids on garbage cans. The removal of rubbish, particularly garden rubbish, is also important.

Rodent-proofing of properties prevents most access to a building. Holes in walls for water and drainage pipes should be sealed to prevent rodent access. Usually cement or metal sheeting prevents access. Inspection of the building to locate entry points is a first step in proofing.

Trapping rats and mice is often done by home-owners and pest control technicians. There are several types of traps from the simple single snap trap to multiple mouse-catching devices which can hold up to 30 mice. The single snap trap may be used with a bait or set unbaited. Suitable foods include bacon, nuts and apple for baited traps. Unbaited traps have a trigger covered with fine pieces of cardboard or sawdust. They are placed in one of the runways, usually at right angles.

Because trapped rats and mice may exude blood, urine or faeces, traps should not be set near food preparation areas. When a rodent is caught its fleas leave the dead body and seek other hosts. It is therefore important to check the traps frequently, and remove rodent and fleas together.

Glueboards, which have an attractant in the glue, may be used in homes, but must be inspected regularly. The rodent must subsequently be killed, and this is not always a favoured task.

THE HOUSE MOUSE
MUS MUSCULUS

head small — fur long and fine

small feet

YOUNG RAT

large head — fur short and fine

large feet

FAECES

House Mouse   Roof Rat

Norway Rat

71

Chemical control

Chemical control methods are preferred by most homeowners, but there are situations where chemicals are not appropriate mainly for safety reasons:

● Dead rodents may die and decompose in inaccessible places.

● Poisoned rodent bodies may be eaten by animal pets.

● Poisoning may not be an option for chemically sensitive people.

● Trapping to mop up the few rodents that have survived a baiting programme may be more economical.

Prior to the 1950s single-dose poisons such as arsenic trioxide and thallium sulphate were in general use against rats and mice. These acute poisons have now been replaced by both single-dose and multiple-dose anticoagulants. These reduce the blood's ability to clot, causing internal haemorrhage and death. The multiple-dose poisons cause the death of rodents in four to eight days of constant feeding, while single-dose anticoagulants cause death in three to seven days after only one feed.

Multiple-dose anticoagulant rodenticides are available as dry baits and tracking powders. To be effective they require continual feeding for four to ten days. The two materials are warfarin ('Ratsak') and coumatetralyl ('Racumin'). It is essential that feeding continues for several consecutive days ensuring the build-up of anticoagulants in the blood, which finally causes death. It is important that the supply of bait is maintained during the baiting period for if the bait is exhausted and not available the baiting period is extended.

The single-dose anticoagulants have the same action as the multiple doses, but continual feeding for several days is not necessary. A single-dose anticoagulant causes death in three to seven days. There are two single-dose rodenticides: brodifacoum ('Talon') and bromadiolone ('Bromakil'). These are available as pellet baits, wax blocks to be used in humid or damp situations, grain baits and drink concentrates.

Both the single-dose and multiple-dose anticoagulants are toxic to other warm-blooded animals and therefore the placement of bait stations must be inaccessible to children, pets, wildlife and livestock. In the event of accidental intake, vitamin K1 is an effective antidote, but accidental poisoning should be referred to a medical practitioner in the case of children, and a veterinary surgeon for animals.

Fumigants such as phosphine are used on farms, where a quick result is required. An operator must be licensed to use these fumigants.

Tracking powders are placed along runways. Rats and mice come into contact with the dust, and ingest the poison when grooming their paws and fur. Those currently available contain single-dose anticoagulants.

RATS & MICE
BAITING AND TRAPPING

1   Obtain an accurate identification of the rodent and a full evaluation of the situation prior to selecting control measures. Different species have differing habits, and dead rodents in inaccessible places can cause problems from odour or blowflies.

2   Read the instructions on the label of the bait container. Follow these instructions carefully.

3   Select the bait or trap which is most suitable for the problem.

4   Place traps on or very near runways, but not near food or accessible to children or other animals.

5   Do not place baits where they are accessible to children or domestic animals.

6   Do not place baits near food, food utensils, or preparation areas.

7   Record the placement of baits/s on paper, so that all bait stations are removed when control is achieved.

8   Do not apply tracking powders near food or utensils where contamination can occur.

9   Wear gloves when handling dead rodents, either poisoned or trapped. Use an aerosol insecticide for the immediate area to eradicate the fleas which leave the dead body.

10   Carefully wash used and contaminated traps, and destroy any bait residue once control is achieved.

CHAPTER 15 CONTROL METHODS

NON-CHEMICAL PROCEDURES

Most household pests can be controlled, or at least significantly reduced, by certain household procedures without the use of chemicals. Even when chemicals are used the following environmental procedures will complement the chemical application so that a house will be almost pest free for long periods.

COCKROACHES

- Eliminate all food particles, mainly in kitchen and cupboards, by using a vacuum cleaner daily.
- Store food in sealed containers, rather than partially open bags.
- Place waste in garbage can in sealed bags.
- Inspect all food containers prior to storage.
- Inspect the motors of refrigerators and other appliances regularly as the warmth generated is attractive.
- Also inspect computers, fax machines and microwave ovens for the same reasons.
- Fill and seal all cracks and crevices in cupboards and walls particularly in and near kitchens.
- Inspect roof cavities, at least annually.

MOSQUITOES

- Keep guttering, rock pools and wherever water is retained dry, or empty every two weeks, to deny larval and pupal stages their breeding places.
- Fit wire screens to reduce the entry of mosquitoes into buildings.

RATS AND MICE

- Keep all doors and other inlets to houses closed, particularly at night.
- Store all foods in sealed containers.
- Close cupboards when not in use.
- Vacuum food-handling and eating facilities daily after the evening meal. Leave no food particles at night.
- Inspect roof cavities and underneath furniture for nest areas.
- Use traps to detect rodents, as well as reducing their population.

CARPET BEETLES

- Vacuum regularly, giving special attention to the edges of rooms and behind and underneath furniture.
- Inspect roof cavities for dead possums and rodents, as these are attractive to the beetles and are food for their larvae.
- Inspect furs of animal origin in wardrobes for these are attractive. Synthetic fabrics are not usually infested, unless soiled by food, urine or other materials.
- Inspect flowers being brought into the house, as carpet beetles are pollen feeders and can be brought in on flowers.

SILVERFISH

- Inspect roof areas, as silverfish often come via ceilings into bathrooms.
- Humidity favours silverfish and attack is often found in book bindings. Air conditioned areas are seldom favoured by silverfish.
- Vacuum roof cavities and book storage facilities annually.

CLOTHES MOTHS

- Treat infested articles of clothing by exposing to sunlight for two to four hours in black plastic bags.
- Vacuum woollen carpets. Attention to woollen covered lounge chairs is also important.
- Inspect walls and the rear areas of furniture to reveal casemaking clothes moth pupal cases. These can be easily removed.
- Inspect woollen articles when purchased.

GRAIN BORERS, WEEVILS ETC.

- Carefully heat-treat small packages of grain products in a microwave oven if their condition is suspect.
- Heat treatment may also be carried out by placing products in black plastic bags and exposing to sunlight for two to four hours.

- Transfer the contents of packages of grain food or products of grain to sealed containers once they are opened.

FLIES

- Dispose of garbage at least weekly in homes and daily in food handling facilities.
- Place all waste food in sealed containers or plastic receptacles before disposal.
- Regularly inspect compost heaps, as these encourage fly breeding.
- Bury all animal faecal material.
- Install screens and 'air curtains' to exclude adult flies from buildings.
- Install electric light traps to reduce adult fly numbers.
- Use sticky fly baits.

FLEAS

- Take care of pets, especially by washing.
- Daily inspect pet bedding materials, and remove flea larvae and pupae by shaking them thoroughly outside.
- Treat pets by either external or oral medication.
- Regularly wash floors, and vacuum or steam-clean carpets (whichever is appropriate).
- Vacuum upholstery of chairs and lounges.
- Heat-treat the contents of the vacuum cleaner in the sun in black plastic, or spray with an aerosol. Do not put away a full bag.
- Vacuum the relevant areas of the house prior to going on holidays, and treat the bag contents before disposal.

BED BUGS

- Carefully inspect second-hand furniture before bringing it into the house.
- Wrap furniture suspected to harbour bed bugs in black plastic and placed in sunlight for two to four hours.
- Inspect suspect clothing or suitcases before bringing it into the house.

ANTS

- Remove all food particles from floors and food preparation surfaces, preferably with a vacuum cleaner.
- Wipe and wash up any spillage of drinks as many ant species are attracted to sugary materials.
- Remove any foodstuffs left outside, particularly where pets are fed, as ants may be attracted to these and then enter the house.

FUNNELWEBS, REDBACKS, WHITETAILED, MOUSE SPIDER AND TRAPDOOR SPIDERS

Remember, most spiders are not pests and are important in our environment, since they reduce the populations of pest insects. However, some are poisonous, and their bites require medical attention.

- Identify spiders found in the house. Kill only those that are regarded as poisonous to humans.
- Take care when using insecticides in gardens, as this can cause funnelweb spiders to enter houses. Wet weather may also cause ground dwelling spiders to enter houses. Ground-dwelling spiders are also often found in swimming pools and can be alive even after hours at the bottom. Do not remove with the hands.
- Carefully inspect all footwear and clothing left outside before dressing, particularly in summer and autumn.
- Inspect all toys, clothing, footwear and other articles left outside at night before bringing them into the house.

LOOSE WEB AND ORBWEB SPIDERS

- Use a vacuum cleaner to remove webs from under eaves and around windows. Release spiders into a more suitable environment.

PARALYSIS TICK

- Inspect and remove excess vegetation.
- Discourage bandicoots, or transfer them to other sites, as they are a natural host of the paralysis tick. Bandicoots are protected animals and must not be hurt or killed.
- Inspect animals daily, with particular attention to head and ears.
- Inspect children. Make them aware of the early symptoms of tick attachment.

LICE

Head lice are the most often encountered. Crab lice also occur, mainly conveyed by sexual contact and a failure to wash underclothing. Body lice are seldom a problem in Australia.

- Avoid contact with others who are infested. Schools often encounter head lice problems so that behaviour of children at home must always be monitored by parents.
- Use a special fine-toothed comb to detect head lice. This also provides some control.
- Bathe and change underclothes regularly to assist in preventing infestation with crab lice where contact with infested persons occurs.
- Wash and iron clothing, bed linen and towels to keep head lice infestations down.

75

MITES

There are several mites which infest humans when they leave their normal hosts. As many people are allergic to mites infestation should be referred to a medical practitioner.

The grass itch mite spreads from vegetation to animals to humans. It is usually detected in the ears of cats as small red clusters.

- Have a veterinary surgeon treat affected cats.

The infestation of scabies mite is related to human hygiene and occurs where this is associated with population density. Personal contact and also infested clothing are the main causes of the spreading of scabies mites.

- Wash and sterilise clothing during the medical treatment of scabies mites. Bed linen and mattresses also require attention.

When bird mites leave nests in roof areas looking for hosts the infestation of humans and animals can occur.

- Remove nests, or young birds before they are old enough to leave the nest.

Mites associated with paper often cause human irritation.

- Inspect paper supplies in offices.

Dust mites occur in many homes.

- Vacuum frequently and immediately dispose of the bag contents.
- Wash clothing and other articles.
- Sterilise bedding in black plastic in sunlight to assist in control.

EUROPEAN WASPS

- Trace wasps to the colony source.
- Do not attempt the removal of the colony. Wasps will sting en masse, resulting in serious illness or death.
- Contact a pest control technician who is trained in this field to eradicate the colony.

PESTICIDES

When a major pest problem occurs in the home the householder usually selects a pest control technician to provide advice and to apply control measures. Some householders choose to carry out their own control measures and often require guidance in the selection and application of a pesticide or non-chemical procedure to eliminate the problem. The following section has been prepared to assist the householders to understand the measures used by the pest control technician, and to guide those who wish to apply the pesticide or other measures themselves.

Often a householder can also perform certain actions which will complement the control measures carried out by the pest control technician so that chemical treatments are reduced to a minimum. These non-chemical procedures were covered in the previous chapter.

PESTICIDE DEVELOPMENT

Before World War II there were very few insecticides used. The naturally occurring pyrethrins, some inorganic minerals such as boron and arsenic, and the fumigant hydrogen cyanide were used during the 1920s and 1930s.

The first *organochlorine* pesticide, dichlorodiphenyltrichloroethane (commonly known as DDT) was developed and used just prior to World War II. Later benzene hexachloride and the cyclodiene organochlorines aldrin, dieldrin, chlordane and heptachlor were used against a variety of insect pests. Soon after that *organophosphorus* insecticides were developed, followed by *carbamates*. Because of their long-term environmental damage the organochlorine insecticides were gradually phased out. Fortunately the less residual organophosphates and carbamates were available to fill the gap created. Later synthetic pyrethroids were developed, and many of these have a low toxicity level for warm-blooded creatures including man.

The insect growth regulators (IGRs) have also been developed during the last twenty years. These have a very specific toxicity, for they are hormones which only affect insects' growth patterns at the time of critical formation, thus eradicating certain species and controlling others.

PESTICIDE GROUPS

INORGANIC INSECTICIDES

These include arsenic trioxide used for termite control, and boric acid and silica aerogels used in cockroach control.

BOTANICALS

These are insecticides obtained from plant materials. They include pyrethrins obtained from the flowers of *Chrysanthemum* sp. They are widely used against pests of the house, garden and stored products. Rotenone is another plant product used for plant pests, and is often referred to as derris dust.

ORGANOPHOSPHATES

This is a large group of insecticides which have been derived from phosphoric acid. These insecticides have largely replaced the more residual organochlorines and have been more acceptable for their shorter residual life of a few weeks to a few months. They are effective against insect pests of the garden, stored products, household pests, crops and for the protection of animals against external parasites.

The selection of the appropriate organophosphate for a particular problem is essential, for they work in different ways. Dichlorvos, a volatile organophosphate enters insects via their respiratory system. Most others such as diazinon, chlorpyrifos, fenthion, maldison, temephos and trichlorphon are oral and contact insecticides, entering through the pests' mouthparts or cuticle (skin).

CARBAMATES

These insecticides have been derived from carbamic acid and are short-term residual insecticides having an effective oral and contact life of a few weeks to a few months. The main use of these insecticides is against household pests. They include bendiocarb, propoxur, carbaryl and methomyl.

SYNTHETIC PYRETHROIDS

These insecticides owe their origins to the natural pyrethrins. The aim was to produce insecticides which were safer to humans and had residual and contact toxicity to pests. The pyrethroids now available are copies of parts of the natural pyrethrin molecule and are used extensively in household pest control. They include products such a permethrin, bioresmethrin, deltamethrin, cyfluthrin, tetramethrin and cypermethrin. They are now a major component of the insecticide market in Australia as most of the safety and environmental pollution requirements for their use have been met.

INSECT GROWTH REGULATORS

These chemicals disrupt the insects' growth patterns and development causing their death usually some months after contact. IGRs disrupt moulting, or in those insects which have pupae it prevents pupal formation or adult emergence. Adult insects are not usually affected by IGRs, but larvae or nymphal stages of insects are affected when they moult in order to pass to the next stage. They have a low toxicity to humans, they are persistent when applied inside a house, and have no odour. However, they take some months to control the target insect. Those in use include hydroprene, methoprene, periproxifen, fenoxycarb and triflumuron.

SELECTING PESTICIDES

Prior to applying any control measures the following procedures are suggested:

1   Collect the pest, and have it identified.

2   Discuss the merits of various treatments, both chemical and non-chemical, with your pest control technician, or carefully consult pesticide label information, and select the method most acceptable to you.

3   If applying the treatment yourself, follow the label information and dilution instructions of the chosen treatment.

4   Ensure that all food is covered, and that pets are removed, during the treatment and for a few hours afterwards.

5   Wear protective clothing, including an appropriate mask and gloves, during the application.

6   Apply the insecticide to the various target areas, described under the particular pest in previous chapters of this book.

7   Allow the insecticide to dry, and the area to be well aerated by opening windows, before entering the treated areas.

8   Make sure that any baits used are not accessible to children and pets.

BUYING PESTICIDES

This tables lists the names of the main active ingredients of some pesticides which are available to the homeowner; some common Trade names; and the pests which the registered products should control and is intended to provide some guidance to homeowners in the selection and purchase of products for specific pests. Always consult the label information regarding the pesticides' suitability for specific pests.

ACTIVE INGREDIENT	COMMON TRADE NAMES	PESTS CONTROLLED
ORGANOPHOSPHATES		
Azamethiphos	Alfacron 500, Residual Insecticide	Cockroaches, flies, spiders
Chlorpyrifos	Deter Insecticide	Cockroaches, spiders, ants, fleas
	Dursban PC Termiticide & Insecticide	Cockroaches, spiders, ants, fleas, silverfish
	Nufarm chlorpyrifos	Cockroaches, spiders, ants, fleas, silverfish, mosquitoes, hide beetles
Diazinon	Neocid 200P Insecticide	Cockroaches, spiders, fleas, ants, silverfish, bed bugs, mosquitoes, carpet beetles
Fenthion	Baytex 550 Insecticide Spray	Cockroaches, fleas, flies, spiders
Pirimiphos-methyl	Actellic Public Health Insecticide	Cockroaches, fleas, flies, mosquitoes
CARBAMATES		
Bendiocarb	Ficam D Insecticide Dust	Cockroaches, fleas, bed bugs, ants, spiders, silverfish, European wasps
	Ficam W Insecticide	Cockroaches, fleas, bed bugs, ants, spiders, silverfish, European wasps
Propoxur	Baygon 200 Cockroach Spray	Cockroaches, ants, bed bugs, flies, mosquitoes, silverfish, European wasps
	Blattanex Professional Insecticide	Cockroaches, ants, bed bugs, clothes moths, carpet beetles, flies, mosquitoes, European wasps
SYNTHETIC PYRETHROIDS		
Bioresmethrin + Piperonyl butoxide	Reslin Thermal Fogging (for misting and fogging)	Moths, flies, biting midges, American cockroach, mosquitoes
Cyfluthrin	Responsar Professional Insecticide	Cockroaches, ants, fleas, flies, silverfish, spiders, bed bugs
Cypermethrin	Cynoff wsb.	Cockroaches, spiders, silverfish, flies, ants, mosquitoes
Deltamethrin	Cislin 10 Residual Insecticide	Cockroaches, ants, spiders, silverfish, flies, mosquitoes
Permethrin	Bolt Commercial Insecticide (aerosol)	Cockroaches, fleas, silverfish, carpet beetles, beetles, spiders, ants, moths
	Coopex Residual Insecticide, Imperator Residual Insecticide	Cockroaches, fleas, silverfish, carpet beetles clothes moths, flies, spiders, mosquitoes, ants, bed bugs
BOTANICAL INSECTICIDES		
Pyrethrins + piperonyl butoxide	Drift (Pestech) (for misting and fogging)	Cockroaches, flies, mosquitoes, ants, silverfish, spiders, pests of stored food
Pyrethrins + piperonyl butoxide	Rudchem PY-40 Formula 40 Insecticide	Cockroaches, flies, mosquitoes, ants, silverfish, spiders, pests of stored food

ACTIVE INGREDIENT	COMMON TRADE NAMES	PESTS CONTROLLED
INSECT GROWTH REGULATORS		
Cyromazine + permethrin	Flego Residual Insecticide	Larval and adult fleas
Fenoxycarb + permethrin	Mortein Flea Bomb	Cockroaches, silverfish, flies, spiders, ants
Hydroprene + permethrin	Protrol Crack and Crevice + permethrin	Cockroaches
Pyriproxyfen	Sumilarv Insect Growth Regulator	Immature fleas, cockroaches
Triflumuron	Starycide 480 SC	Immature fleas, cockroaches
OTHER CHEMICALS		
Hydramethylnon	Maxforce Cockroach Control System — Household Insecticide Baits	Cockroaches (There are also formulations for ants.)
RODENTICIDES		
Coumatetralyl	Racumin Mouse and Rat Rodenticide	Tracking powder & baits
Warfarin	Rentokil Bait for Rats and Mice	Ready-to-use baits
	Ratsak Ready to Use	Ready-to-use baits
Brodifacoum	Talon G. Rodenticide Pellets	Ready-to-use baits
	Talon Rodenticide, All Weather Rodenticide Wax Block	Wax block containing food plus rodenticide for rodents to feed on
Bromadiolone	Bromakil Super Rat Bait	Ready-to-use grain-based bait
	Bromakil Super Rat Blocks	Ready-to-use wax blocks contain food and rodenticide
	Rentokil Bromakil Super Rat Bait	Ready-to-use grain and rodenticide
	Rentokil Super Rat Blocks,	Ready-to-use wax blocks containing food & rodenticide

LABEL INFORMATION

- The product label must contain the specific uses for which the pesticide is intended. Note that label data changes from time to time, and may be different for each state in Australia.

- There may be several different formulations for a particular product. Each formulation may have specific uses and these uses would appear on the product label. If the intended use for the pesticide does not appear on the label it should not be used.

- The label will also contain dilution instructions where applicable, and these should be followed carefully. Some products such as aerosols are not diluted.

- The label will also provide basic information on procedures to follow in case of accidental contact with or poisoning by the pesticide.

PESTICIDE SUPPLIERS

While most pesticides can be obtained from hardware stores, and some plant nurseries, there are companies which supply the pest control industry. Some products such as those for termite soil barrier treatments, fumigants and more toxic chemicals can only be supplied to licensed pest control technicians.

The following companies supply a range of pesticides, many of which have been mentioned in previous chapters of this publication, and have branches in several states of Australia:

- Chemical Enterprises Pty. Ltd.
- Garrards Pesticides Pty. Ltd.
- Globe Australia Pty. Ltd.

INDEX

FURTHER READING

Cornwell, P.B. (1979) *Pest Control in Buildings*, Rentokil Ltd, Sussex, England.

Hadlington, P. and Gerozisis, J. (1995) *Urban Pest Control in Australia*, UNSW Press, Sydney.

Hadlington, P. and Johnston, J. (1987) *An Introduction to Australian Insects*, UNSW Press, Sydney.

Hadlington, P. and Marsden, C. (1998) *Termites and Borers: A Homeowner's Guide to Detection and Control*, UNSW Press, Sydney.

Naumann, Ian (1993) *CSIRO Handbook of Australian Insect Names*, CSIRO, Australia.

Simon-Brunet, B. (1994) *The Silken Web*, Reed Books, Sydney.